Asteroseismology for the Nonspecialist

Online at: https://doi.org/10.1088/2514-3433/ae03a0

AMERICAN ASTRONOMICAL SOCIETY IOP | ebooks™

About the program:

AAS-IOP Astronomy ebooks is the official book program of the American Astronomical Society (AAS) and aims to share in depth the most fascinating areas of astronomy, astrophysics, solar physics and planetary science. The program includes publications in the following topics:

- GALAXIES AND COSMOLOGY
- INTERSTELLAR MATTER AND THE LOCAL UNIVERSE
- STARS AND STELLAR PHYSICS
- EDUCATION, OUTREACH, AND HERITAGE
- HIGH-ENERGY PHENOMENA AND FUNDAMENTAL PHYSICS
- THE SUN AND THE HELIOSPHERE
- THE SOLAR SYSTEM, EXOPLANETS, AND ASTROBIOLOGY
- LABORATORY ASTROPHYSICS, INSTRUMENTATION, SOFTWARE, AND DATA

Books in the program range in level from short introductory texts on fast-moving areas, graduate and upper-level undergraduate textbooks, research monographs, and practical handbooks.

For a complete list of published and forthcoming titles, please visit iopscience.org/books/aas.

About the American Astronomical Society

The American Astronomical Society (aas.org), established 1899, is the major organization of professional astronomers in North America. The membership (~7,000) also includes physicists, mathematicians, geologists, engineers, and others whose research interests lie within the broad spectrum of subjects now comprising the contemporary astronomical sciences. The mission of the Society is to enhance and share humanity's scientific understanding of the universe.

Editorial Advisory Board

Asteroseismology for the Nonspecialist

Derek L Buzasi
University of Chicago, Chicago, IL, USA

IOP Publishing, Bristol, UK

ISBN 978-0-7503-5633-6 (ebook)
ISBN 978-0-7503-5631-2 (print)
ISBN 978-0-7503-5634-3 (myPrint)
ISBN 978-0-7503-5632-9 (mobi)

DOI 10.1088/2514-3433/ae03a0

Version: 20251101

AAS–IOP Astronomy
ISSN 2514-3433 (online)
ISSN 2515-141X (print)

British Library Cataloguing-in-Publication Data: A catalogue record for this book is available from the British Library.

Published by IOP Publishing, wholly owned by The Institute of Physics, London

IOP Publishing, No.2 The Distillery, Glassfields, Avon Street, Bristol, BS2 0GR, UK

US Office: IOP Publishing, Inc., 190 North Independence Mall West, Suite 601, Philadelphia, PA 19106, USA

For Heather and Grant, who tolerated with love all the late nights and time away from home over all the years, and supported me when I most needed it. Truly my favorite and most-loved people.

Contents

Preface

Why this book? After all, there are wonderful introductions to Asteroseismology out there already, including Aerts et al. (2010), Basu & Chaplin (2018), Pallé & Esteban (2014), as well as a range of magisterial review articles, including Aerts (2021), Hekker & Christensen-Dalsgaard (2017), Garcia & Stello (2018), García & Ballot (2019), Kurtz (2022), Bowman (2020), Di Mauro (2016), and many others.

My motivation for writing this book derives from a recognition that the field has changed. Since the launch of *Kepler*, it has transitioned from a thing people **do** to a thing people **use**, and in the process it has touched essentially every aspect of stellar astrophysics (and beyond). This is a great thing, and speaks to the maturity and power of the field, but it also means that there's an increasing need for users of asteroseismic results to understand better where those results came from, how they were derived, and what their limitations might be. Those users constitute the audience for this book, though of course I hope it might also be useful to students at the advanced undergraduate or beginning graduate level, or to astronomers in other areas interested in having a go at asteroseismology but perhaps not wanting to start with a 700 page introduction.[1]

About 25 years ago, I became an asteroseismologist by accident. I quickly learned that the most wonderful part of the field is as much the people as the science, and I want to express my appreciation for those who were there to show me the ropes when I first appeared, including Conny Aerts, Vichi Antoci, Thierry Appourchaux, Sarbani Basu, Tim Bedding, Tim Brown, Hans Bruntt, Bill Chaplin, Jørgen Christensen-Dalsgaard, Margarida Cunha, Jan Cuypers, Pierre Demarque, Yvonne Elsworth, Rafa Garcia, Joyce Guzik, Saskia Hekker, Steve Kawaler, Hans Kjeldsen, Don Kurtz, Travis Metcalfe, Maria Pia Di Mauro, Marc Pinsonneault, Jesper Schou, Juan Carlos Suárez, Rich Townsend, Regner Trampedach, Werner Weiss, and Konstanze Zwintz. Thanks more than I can say to all of you.

About 25 years ago, I became an asteroseismologist by accident. I've tried to write the book I would have wanted then.

References

Aerts, C. 2021, Rev. Mod. Phys., 93, 015001

Aerts, C., Christensen-Dalsgaard, J., & Kurtz, D. W. 2010, Asteroseismology (Dordrecht: Springer)

Basu, S., & Chaplin, W. J. 2018, Asteroseismic Data Analysis. Foundations and Techniques (Princeton, NJ: Princeton Univ. Press)

Bowman, D. M. 2020, FrASS, 7, 70

Di Mauro, M. P. 2016, Frontier Research in Astrophysics II (FRAPWS2016) (Trieste: PoS) 29

[1] But those books and reviews I listed? Eventually, you should read them all.

García, R. A., & Ballot, J. 2019, LRSP, 16, 4
Garcia, R. A., & Stello, D. 2018, arXiv:1801.08377
Hekker, S., & Christensen-Dalsgaard, J. 2017, A&AR, 25, 1
Kurtz, D. W. 2022, ARA&A, 60, 31
Pallé, P., & Esteban, C. 2014, Asteroseismology, Canary Islands Winter School of Astrophysics (Cambridge: Cambridge Univ. Press)

About the Author

Derek L Buzasi

Derek received his undergraduate degree in physics from the University of Chicago, and his PhD in astronomy from Penn State University. Since then, he has worked at a variety of institutions, including the National Center for Atmospheric Research, Johns Hopkins University, the California Institute of Technology, and the University of California at Berkeley, and on a number of major spacecraft and instrument teams, including the HST Cosmic Origins Spectrograph, NASA's planet-finding Kepler mission, and the Wide-Field Infrared Explorer satellite (WIRE), which he used to perform the first asteroseismology from space.

Derek's research interests include almost anything having to do with stars. He began by studying various aspects of stellar (and solar) activity, such as spots, flares, and winds, and has done both observations and theoretical work, including radiative transfer modeling and magnetohydrodynamic models of stellar flux tubes. More recently, he has moved from primarily studying stellar atmospheres and environments to studying stellar interiors and convection through the use of asteroseismology.

He is currently Senior Instructional Professor in the Department of Astronomy and Astrophysics at the University of Chicago.

Asteroseismology for the Nonspecialist

Derek L Buzasi

Chapter 1

A Brief Review of Stellar Structure and Evolution

Stars are among the most important objects in the Universe. They comprise much of the baryonic mass, drive both chemical and dynamical evolution, and act as tracers of structure on both galactic and larger scales. Stars also serve as hosts to planets, and thus are intimately associated with life.

Understanding the internal structure and evolution of stars has been an ongoing enterprise since at least the later part of the nineteenth century, but the first true understanding arose when knowledge of nuclear physics was integrated into their study in the 1940s, allowing a window into the physical processes responsible for energy production in stars during the majority of their lifetimes. More recently, improvements in our understanding of processes as diverse as equations of state, convection, opacity, and stellar magnetism have led to computer models which approach more and more closely the physical parameters we observe in stars.

Historically, stellar observations began with gross properties, including mass, radius, luminosity, and temperature. Calibrations were derived from clusters and eclipsing binaries, and uncertainties in these parameter values were relatively large, 10%–20% or even more. Approaches to stellar modeling then involve construction of a computer model incorporating as complete a set of detailed physics as possible, and applying surface boundary conditions to match the observables.

Figure 1.1 shows a Hertzsprung–Russell diagram based on observational data from the *Gaia* spacecraft (Gaia Collaboration et al. 2023). The vertical axis is given in *Gaia* absolute magnitudes M_{G}, which serve as a proxy for luminosity and on which scale the Sun has $M_{\mathrm{G}} = +4.67$, while the horizontal axis is in the magnitude difference between the *Gaia* blue and red filters, $G_{\mathrm{BP}} - G_{\mathrm{RP}}$, a proxy for temperature; here the solar value is $G_{\mathrm{BP},\odot} - G_{\mathrm{RP},\odot} = 0.82$ (Casagrande, 2018). Note that astronomical magnitude is a logarithmic measure in which magnitude $m = -2.5\log_{10}F + \mathrm{zp}$, where zp is a zero-point offset appropriate to the band in which the flux F is measured (Andrae et al. 2018).

doi:10.1088/2514-3433/ae03a0ch1

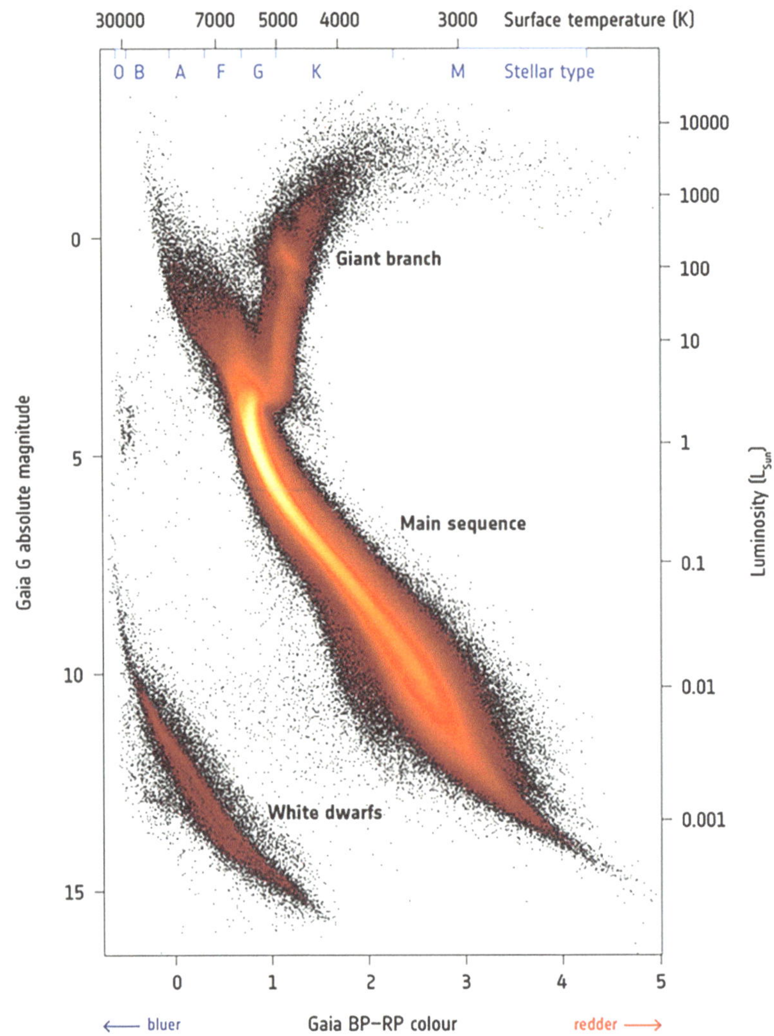

Figure 1.1. A color–magnitude diagram based on data from *Gaia* Data Release 2. Luminosity increases upwards, while temperature increases to the left. The color indicates the number density of stars, so yellow stars are the most common in the sample. The two strips running from top left to bottom right are (on top) the main sequence, where stars are fusing hydrogen to helium in their cores, and (below) the white dwarf sequence, representing the end state of low and intermediate-mass stars. The base of the red giant branch is also visible splitting off from the main sequence and moving upwards and to the right.

As is readily apparent, the majority of stars lie on the main sequence, whose stars are all fusing hydrogen to helium in their cores. The main sequence has a finite width, due primarily to variations in initial composition (metallicity, or [Fe/H]) or changes in core composition as fusion continues through the lifetime of the star. At the cool red end, M dwarf stars have effective temperatures[1] as low as $T_e \sim 2300$K

[1] This is the temperature of a blackbody with equivalent luminosity and surface area.

and radii as low as $\sim$0.1 $R_\odot$, while at the upper end main sequence stars can have radii as large as $\sim$14 $R_\odot$ (or more) and effective temperatures as large as $T_e \sim 45{,}000$K (Pecaut & Mamajek 2013).[2] These radii and effective temperatures imply that the stellar luminosities on the main sequence vary from $\sim 2.8 \times 10^{-4}$ $L_\odot$ at the low end all the way up to $\sim 6.6 \times 10^5$ $L_\odot$ at the top of the upper main sequence. Note that while the bottom of the main sequence is visible in this observational HR diagram, the upper end is not—very hot and bright stars are extremely rare!

The masses of stars vary widely as well, though not by as much as their luminosities do. As we will shortly see, the internal temperature of stars is a function of their mass, so the lower-mass limit on the main sequence is set at $\sim$0.078 $M_\odot$ because stars less massive than this don't reach internal temperatures high enough to support hydrogen fusion.[3] The upper mass limit is less clear, but a main sequence star with $T_e = 45,000$ K has a mass of $\sim$59 $M_\odot$, so we will take that as a convenient upper bound, since stars more massive are exceedingly rare. Since both mass and luminosity increase as we move up the main sequence, this implies a relationship; the simplest relationship is a power-law, and taking just the top and bottom limits implies one of the form

$$L \propto M^{3.26} \tag{1.1}$$

Of course the ripples in the shape of the main sequence show that this is oversimplified, and the best index varies somewhat over different parts of the main sequence. Broadly speaking, though, the reason for the existence of such a simple relationship is simply *because* all the stars on the main sequence share a similar mechanism for producing energy: hydrogen fusion. Note that the existence of a mass–luminosity relationship also implies something about the lifetimes of different stars on the main sequence. Stars at the top have more mass available for fusion, but they also fuse at a much higher rate (hence their much larger luminosities), so other things being equal, they will have shorter main sequence lifetimes. In fact, simple scaling arguments would suggest that an upper main sequence star will have a main sequence lifetime about 3.2×10^6 times shorter than a star at the bottom of the main sequence. Reality is a bit more complicated, but the basic idea still holds up.

There are a few other important features visible in Figure 1.1. First, there is a handful of objects visible at the very bottom of the diagram, even below the $\sim$0.078 $M_\odot$ hydrogen-burning limit. These are *brown dwarfs*, which though incapable of fusing hydrogen to helium, *are* capable of fusing deuterium (^{2}H), an isotope of hydrogen which fuses a bit more easily. Second, there's a sort of parallel sequence to the left of the main sequence and below it; these objects are roughly 10 magnitudes (a factor of 10^4) fainter than main sequence stars with the same effective temperature, implying that they have 10^4 times less radiating surface area, or radii 100 times smaller, than main sequence stars. These are the *white dwarfs*, which are the end

[2] An updated version of this incredibly useful table can be found online at http://www.pas.rochester.edu/emamajek/EEM_dwarf_UBVIJHK_colors_Teff.txt.

[3] The exact value depends a bit on details of the star's composition.

products of the evolution of most of the stars on the main sequence. They are no longer undergoing nuclear fusion of any kind, so their luminosity is due almost entirely to the fact that they are hot, and take a long time to cool off because they are so small. In fact the white dwarf sequence is an evolutionary sequence, in that white dwarfs start off at the top of it and progress downwards as they cool.[4] Finally, there's a kind of spur off the main sequence, representing stars which are cooler than the Sun, but considerably more luminous than main sequence stars of the same effective temperature, implying they have larger radii. These are the *red giants*; the ones visible in this HR diagram have radii ~10–20 times larger than stars on the main sequence, but giant stars can get considerably larger than this!

An important question to address is: how do we know the effective temperatures, radii, luminosities, and masses of these stars? How do we *calibrate* the HR diagram? This is important because that calibration provides an upper bound to how precisely we can know the physical parameters of *any* star.

The most important parameter to measure is distance, and the most direct way to measure distance is by measuring a star's *parallax* angle, essentially half of the apparent shift in the star's position as the Earth moves 180° in its orbit around the Sun. The distance to the star in parsecs is then just the inverse of the parallax angle, so

$$d = \frac{1}{\tilde{\omega}} \tag{1.2}$$

The primary purpose of the *Gaia* mission was to measure parallaxes for billions of stars in our Galaxy, and the precision with which it did so is summarized in Figure 1.2. The sweet spot is at about $m_{\mathrm{G}} = 12$, because fainter stars don't deliver enough photons to the detector, while brighter ones can saturate it. The Sun has a *Gaia* absolute magnitude of $M_{\mathrm{G}} = 4.67$, so the Sun would have $m_{\mathrm{G}} = 12$ at a distance of about 8.5 kpc, at which point it would have $\tilde{\omega} \approx 1.18$ mas. Since the best *Gaia* parallaxes have uncertainties of ~ 0.007 mas, this implies a best-case distance uncertainty of about 0.6%.[5] Luminosities are then derived from measured fluxes combined with distances, with some additional uncertainty derived from the issue of conversion from the flux measured in various bandpasses to the bolometric flux.

We can also measure the angular diameter of stars using optical interferometry (see Baines et al. (2023) as an example among many). Once an interferometric angular diameter θ_{LD} is determined, where the LD indicates that stellar limb darkening must be taken into account, it can be combined with the distance to obtain a physical radius. The resulting typical radius uncertainties are around 1%, with best-case measurements a few times better than that. The best measurements are for giant stars, which have larger radii. Effective temperatures can be derived directly from angular radii and bolometric flux, where

[4] Once upon a time, before stars were at all well understood, the main sequence was believed to operate in the same way.

[5] This will continue to improve with future data releases, potentially reaching uncertainties of only ~ 0.001 mas $= 1\mu$ as.

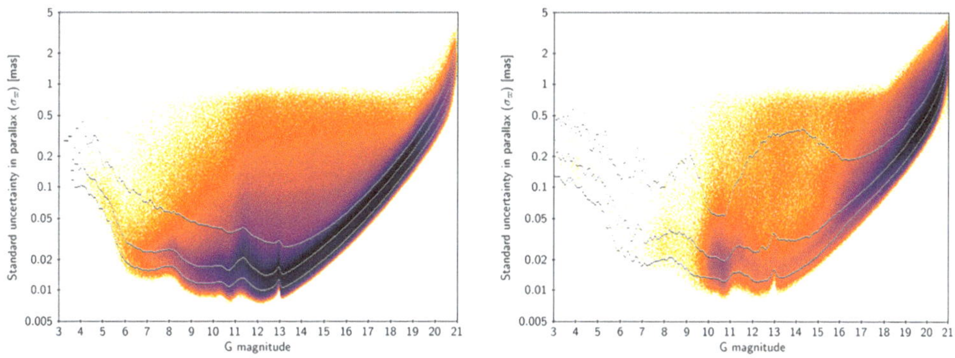

Figure 1.2. Gaia uncertainties in parallax versus magnitude, for all sources brighter than Gmag = 11.5 and a random sample of fainter sources. The left panel shows solutions based on five parameters, while those in the right panel are based on six. The color scale indicates density of data points, and curves shows the 10th, 50th, and 90th percentiles of the distribution. Reprinted with permission from Lindegren et al. (2021).

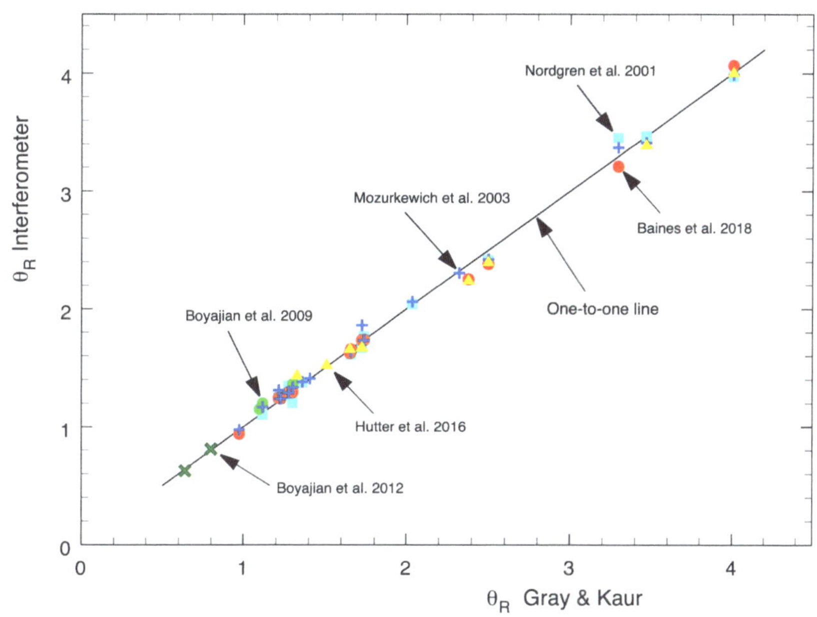

Figure 1.3. Interferometer-determined radii compared to spectroscopic values. The solid line indicates the 1:1 ratio. Reprinted with permission from Gray & Kaur (2019).

$$F_{\text{bol}} = \theta_{\text{LD}}^2 \sigma T_{\text{e}}^4, \tag{1.3}$$

or derived from spectroscopy. In the latter case, radius is inferred from distance and effective temperature, and in the best cases these spectroscopically inferred radii may have comparable uncertainties to those derived from interferometry (Figure 1.3).

An alternative approach is the use of eclipsing binary stars. If both photometric light curves and radial velocity light curves are available, the mass, radius, and

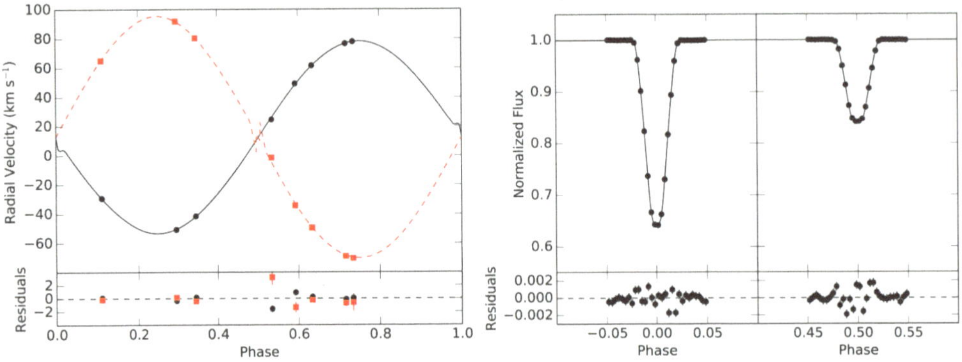

Figure 1.4. The right panel shows radial velocities and the best-fit model for the eclipsing binary KIC 873624, with residuals shown below. The solid black curve shows the radial velocity of the primary star, while the dashed red curve shows that of the secondary component. The left panel shows the binned Kepler light curve, phase-folded at the orbital period, after removing any out-of-eclipse trends. The solid curve is the best-fit model. Reprinted with permission from Fetherolf et al. (2019).

effective temperature of each component can be derived. The photometry shows the eclipses and allows precise measurements of the orbital period and the relative sizes and effective temperatures of the two components, while radial velocity light curves allow the total mass of the system to be accurately decomposed into contributions from each component. An example is shown in Figure 1.4, which illustrates input data for the eclipsing binary KIC 873624 from Kepler photometry (Borucki et al. 2010) and spectroscopy from the Hobby–Eberly Telescope (Ramsey et al. 1998), along with model fits for both. The resulting uncertainty is of order $\sim$1% for mass, $\sim$0.5% for radius, and $\sim$0.7% for effective temperature, which is a reasonable representation of the limiting precision available. For stellar masses, eclipsing binaries are the gold standard.

1.1 Internal Stellar Conditions

It's worthwhile to get an overall sense of the physical conditions inside a star, using the Sun as our baseline. We can measure the mass of our star to be $M_\odot = 2 \times 10^{30}$ kg using Kepler's third law, and the radius we've determined from geometrical considerations to be $R_\odot = 7 \times 10^8$ m. Combining these two parameters gives the average density, $\rho_\odot = 1400$ kg m^{-3}. Main sequence stars in general range from 0.08 – 100 $M_\odot$ and 0.1–15 $R_\odot$, implying a density range of 0.03–80 times solar, with lower-mass stars being denser on average.

The internal pressure and temperature vary with radius, reaching maximal values at the core, but we can estimate representative internal values. For pressure, dimensional arguments suggest

$$P = \frac{\text{Force}}{\text{Area}} = \frac{GM^2/R^2}{4\pi R^2} = \frac{GM^2}{4\pi R^4} \sim 8.8 \times 10^{13}\ \text{Pa} \tag{1.4}$$

and we can use the ideal gas law to derive a corresponding temperature

$$P = \frac{\rho k T}{\mu m_{\rm H}} \rightarrow T = \frac{\mu m_{\rm H} P}{\rho k} = 3.8 \text{ MK} = 3.8\left(\frac{M/M_{\odot}}{R/R_{\odot}}\right) \text{MK} \tag{1.5}$$

Here MK is 10^6K and we've assumed that the solar interior is composed entirely of fully ionized hydrogen, so each hydrogen atom contributes two particles, a proton and an electron, so the mean molecular weight $\mu = 0.5$. This is a reasonable estimate; more precise models imply a central solar temperature a few times this value. This implies a range of typical internal temperatures along the main sequence ranging from roughly 3–20 MK.

1.2 Stellar Timescales

Although we will begin below by considering static, unchanging stars, one helpful aid for understanding stellar behavior derives from considering the timescales on which change can occur. We will work from the longest timescales to the shortest.

Most stars shine by nuclear fusion processes that transform their internal chemical structure, releasing energy in the process.[6] For main sequence stars, the fusion process is converting hydrogen into helium. The precise process varies (CNO cycle and p–p chain are the most common), but the net result is

$$4\text{H}^1 \longrightarrow \text{He}^4 + \nu + \gamma + \text{Energy} \tag{1.6}$$

For the p–p chain, the energy liberated is 26.2 MeV per helium nucleus, which corresponds to the conversion of approximately 0.007 of the rest mass energy of the initial hydrogen. Accordingly, in the best-case scenario in which all of the mass M of a star composed of pure hydrogen is converted to helium, the energy released is $0.007Mc^2$. If the star shines with luminosity L, this implies a *nuclear* timescale of

$$\tau_{\rm nuc} = \frac{0.007Mc^2}{L} \tag{1.7}$$

In the case of the Sun, this corresponds to $\sim 10^{11}$ yr, which is really an upper limit for the age of the Sun because not all of the hydrogen in the Sun is in the core and accessible for fusing and because the luminosity of the Sun will increase as it ages and evolves both on the main sequence and (especially) along the giant branch.

Prior to the early twentieth century and our understanding of nuclear fusion as the primary energy source for stars, energy production was thought to be due to gravitational contraction. This is still an important energy generation mechanism during parts of the life of a star, such as when it is a protostar contracting toward the main sequence. The timescale here is known as the *Kelvin–Helmholtz* time $\tau_{\rm KH}$.[7] Within factors of two that we don't really care about, the virial theorem tells us that

[6] An important exception is pre-main-sequence stars, which release gravitational energy as they contract, a mechanism which also contributes significant energy at some later stages of stellar evolution, particularly for high-mass stars.

[7] After the two famous 19th century physicists; more prosaically it is sometimes called the *thermal* timescale.

the stored thermal energy in a star is roughly equal to the gravitational potential energy, so we have

$$\tau_{\mathrm{KH}} = \frac{GM^2/R}{L} \tag{1.8}$$

For the Sun, this amounts to $\sim 3 \times 10^7$ yr, a few orders of magnitude less than the nuclear timescale.

The fastest timescale is the so-called *free-fall* or *dynamical timescale*, which represents the fastest time over which a substantial fraction of the star can react to a dynamical perturbation. For example, in the most extreme case, if the pressure supporting the star against gravity were suddenly removed.[8] In that case the outer layers would begin to free-fall toward the center, and they'd reach that point after a time

$$\tau_{\mathrm{ff}} \simeq \left(\frac{2R^3}{GM}\right)^{1/2} \tag{1.9}$$

This is also roughly the orbital period for an object orbiting the star at its surface. In the case of the Sun, the free-fall time is roughly $\tau_{\mathrm{ff}} = 2300$s, a bit less than 40 min. Note that this is essentially the fastest time period over which the *entire star* can change dynamically; smaller portions of the star can adjust faster. As we shall see, this timescale is suggestively close to the period at which the Sun oscillates. Along the main sequence such a scaling would suggest oscillation periods ranging from a bit less than the solar value at the bottom of the main sequence, up to a few hours at the top, while on the giant branch we would expect periods to be much longer.

1.3 Equations of Stellar Structure

Modern approaches to modeling the interiors of stars date back to Emden (1927), Chandrasekhar (1939), Eddington (1926), and Schwarzschild (1958). These approaches treated stars as spherically symmetric objects that are in equilibrium. The first assumption obviously ignores the effects of rotation, tides, and pulsations, while the latter assumes that evolutionary changes to global quantities such as mass and luminosity are at least slow compared to the relevant characteristic timescales of the star. Under these circumstances, it is relatively straightforward to write down a set of equations of stellar structure.

The first and simplest equation simply expresses the conservation of mass, in the sense that, if we divide the star up into concentric spherical shells, like an onion, the total mass of the star must equal the sum of the masses of the individual shells. As the thicknesses of the shells is allowed to become infinitesimal, this can be expressed differentially, so that

$$dM = 4\pi r^2 \rho(r) dr \tag{1.10}$$

[8] Posit a magical switch taking the temperature of the gas to approximately 0 K.

Equivalently, we can write this in integral form as

$$M_r = \int_0^R 4\pi r^2 \rho dr \tag{1.11}$$

where R is the stellar radius and M_r is a useful shorthand expression meaning "the mass internal to radius r."[9]

Next we enforce the condition of hydrostatic equilibrium, that the net force on any small parcel of gas within the star vanishes. Clearly this is not *instantaneously* true; even a cursory glance at the outer layers of the Sun reveal them to be in continuous and time-varying motion, and internal convection is generally a significant energy transport mechanism within stars. However, on average stars don't change much on convective timescales, so we consider the static case, keeping in mind that it is only correct in an average sense.

Consider a small cylindrical volume with cross-sectional area σ, oriented radially and located at some distance r from the center of the star. Over a small change in radius dr, the change in the pressure force (which increases inwards) must be balanced by the change in the inward force of gravity, or

$$dP = -\rho g dr = -\rho \frac{GM_r}{r^2} dr \tag{1.12}$$

The one-dimensional nature of our mathematical statement of the problem is clear here, because we ignore the other two positional coordinates θ and ϕ due to the radial symmetry of the problem.

Of course, as noted above, stars produce energy, primarily by nuclear fusion in the deep interior, but also by gravitational contraction at some portions of their life cycles. For now, in keeping with our quasi-static idea of a star, we consider only nuclear energy sources, whose energy output changes only slowly while stars are on the main sequence. In addition to conservation of mass, we can consider conservation of energy, which can be thought of in a similar fashion: the sum of the energy released in each concentric spherical shell must equal the energy released by the star as a whole: its luminosity L.

$$dL = 4\pi r^2 \rho \varepsilon dr \tag{1.13}$$

Here the simple term ε represents the energy release per unit time per unit mass, and of course is generally a function of both composition and the various thermodynamic quantities. Once again, we can equivalently write this in integral form as

$$L = \int_0^R 4\pi r^2 \rho \varepsilon dr \tag{1.14}$$

Energy generation in the quasi-static case is generally through nuclear fusion, which on the main sequence involves the conversion of hydrogen into helium. There is a small difference in mass between four hydrogen nuclei and one helium nucleus, the

[9] Later we'll use L_r which has the same application to luminosity.

so-called *mass defect*, and it is that mass difference Δm which is converted into energy via $E = \Delta mc^2$. It is important to note that while the *net* reaction may be as shown in Equation (1.6), this net reaction is constructed through several interconnected reaction chains. In the Sun, the dominant chain is the *p–p chain*

$$\begin{aligned} {}^1_1\mathrm{H} + {}^1_1\mathrm{H} &\rightarrow {}^2_1\mathrm{H} + \mathrm{e}^+ + \nu_\mathrm{e} + 0.42\,\mathrm{MeV} \\ {}^2_1\mathrm{H} + {}^1_1\mathrm{H} &\rightarrow {}^3_2\mathrm{He} + \gamma + 5.493\,\mathrm{MeV} \\ {}^3_2\mathrm{He} + {}^3_2\mathrm{He} &\rightarrow {}^4_2\mathrm{He} + 2{}^1_1\mathrm{H} + 12.589\,\mathrm{MeV} \end{aligned} \tag{1.15}$$

Here two each of the first two steps must take place for each of the final step. Positrons created during steps in the chain rapidly annihilate with free electrons in the surrounding medium, producing additional energy, while neutrinos produced have such a low interaction cross-section that they escape the star, taking their energy with them.

Strictly speaking, this is the *p–p I* chain, which is the dominant variant at temperatures similar to those found in the core of the Sun. There are *p–p II* and *p–p III* variants which coexist with *p–p I*, but only become significant contributors at higher temperatures. These simple net reactions occur via complex reaction chains because the probability of multi-component collisions happening simultaneously (which would be required for **four** particles in the case of Equation (1.6)!) is extremely small.

Temperature matters, because in order for fusion reactions to occur the participating particles must be brought close enough that the strong nuclear force can mediate the interaction between them, a distance on the order of 1 fm. This is complicated by the fact that the protons both have positive charges, so experience Coulomb repulsion, or a *Coulomb barrier*, which for particles of charges Z_1 and Z_2 is roughly $Z_1 Z_2$ MeV. Since $\frac{3}{2}kT = 1$ MeV for $T \sim 10^{10}$K, even deep in the core few particles are actually moving fast enough to overcome the barrier. This both limits the reaction rate and causes it to be highly sensitive to temperature.

At modestly higher temperature, such as found in the cores of stars on the upper main sequence, energy generation is dominated by the *CNO cycle*, which has the same net reaction as does the pp chain, but which looks like this:

$$\begin{aligned} {}^{12}_6\mathrm{C} + {}^1_1\mathrm{H} &\rightarrow {}^{13}_7\mathrm{N} + \gamma + 1.95\,\mathrm{MeV} \\ {}^{13}_7\mathrm{N} &\rightarrow {}^{13}_6\mathrm{C} + \mathrm{e}^+ + \nu_\mathrm{e} + \gamma + 1.20\,\mathrm{MeV} \\ {}^{13}_6\mathrm{C} + {}^1_1\mathrm{H} &\rightarrow {}^{14}_7\mathrm{N} + \gamma + 7.54\,\mathrm{MeV} \\ {}^{14}_7\mathrm{N} + {}^1_1\mathrm{H} &\rightarrow {}^{15}_8\mathrm{O} + \gamma + 7.35\,\mathrm{MeV} \\ {}^{15}_8\mathrm{O} &\rightarrow {}^{15}_7\mathrm{N} + \mathrm{e}^+ + \nu_\mathrm{e} + 1.13\,\mathrm{MeV} \\ {}^{15}_7\mathrm{N} + {}^1_1\mathrm{H} &\rightarrow {}^{12}_6\mathrm{C} + {}^4_2\mathrm{He} + 4.96\,\mathrm{MeV} \end{aligned} \tag{1.16}$$

Again, there are variant CNO chains as there were p–p ones, but this version is the dominant contributor. At even higher temperatures, He fusion becomes possible, and is driven by the *triple alpha process*, a simple chain involving three helium nuclei in two reactions,

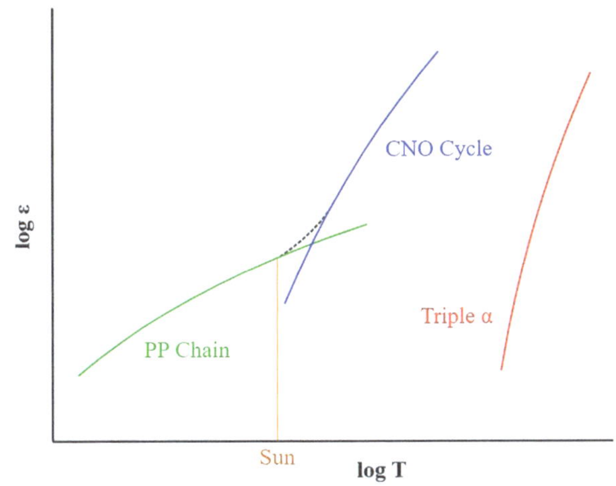

Figure 1.5. A schematic representation of the different sources of nuclear energy inside stars on and relatively close to the main sequence. For stars roughly solar mass and below, the dominant source is the pp chain, while more massive stars on the main sequence make use of the CNO cycle. The triple-α process represents the dominant pathway to helium fusion inside stars where the internal temperature approaches 100 MK.

$$\begin{aligned} {}^4_2\text{He} + {}^4_2\text{He} &\rightarrow {}^8_4\text{Be} + (-0.0918)\ \text{MeV} \\ {}^8_4\text{Be} + {}^4_2\text{He} &\rightarrow {}^{12}_6\text{C} + \gamma + 7.367\ \text{MeV} \end{aligned} \tag{1.17}$$

The first step of this process produces an unstable product, ${}^8_4$Be, but the decay time for this nucleus is long compared to the typical time between interactions with another ${}^4_2$He nucleus. Figure 1.5 schematically shows the energy generation rates from all three processes discussed above.

At higher temperatures, fusion of higher-Z nuclei becomes possible, and these reactions become important in the later stages of the evolution of higher-mass stars. However, we will disregard these reactions, since this book focuses primarily on main-sequence stars and red giants undergoing H and He fusion only.

Once energy is generated, it is transported from one part of the star to another, moving in the direction of decreasing temperature T and thus flowing on average outwards to be released into space. Broadly speaking, there are two transport mechanisms we concern ourselves with, radiation and convection. In the first case, the energy transport rate is determined by both the temperature gradient dT/dr, and the *opacity* κ of the local gas, while convective energy transport is both more complex and more poorly understood.

1.4 Radiation Transfer and Opacity

Adopt the plane-parallel approximation, in which we look at a small enough piece of each spherical shell that we can treat it as flat, and consider the resulting slab of material inside a star. For simplicity, take the radius r at which the slab lies to be large enough to lie outside the nuclear fusion region, so we can neglect energy generation ε. In that case, a beam of monochromatic radiation transiting the slab loses intensity I as it does so, so that

$$\frac{dI}{I} = -\kappa\rho dx \tag{1.18}$$

which integrates to become

$$I = I(0)\exp\left(-\int \kappa\rho dx\right) = I(0)e^{-\tau} \tag{1.19}$$

Here the quantity τ is the *optical depth,* a dimensionless quantity where one optical depth τ is the distance over which the intensity of incoming radiation is reduced by the factor $1/e$. We can express it in terms of physical quantities as

$$\tau = \kappa\rho x \tag{1.20}$$

where x is the path length, ρ is the density, and κ is the opacity, with units of $m^2\ kg^{-1}$.[10]

Despite the name, opacity may be caused by processes that both absorb light and scatter it. Intuitively, it may be helpful to think of the gas as consisting of many tiny spherical absorbers, each with *cross section* σ; the opacity κ then consists of the total cross section for each kilogram of gas. For a sense of scale, the Thomson cross-section for nonrelativistic scattering of light from an electron is $\sigma_T = 6.625 \times 10^{-29} m^2$.

If we allow the slab of material to both emit and absorb radiation, things get slightly more complex, because emission from deeper layers is absorbed by shallower layers, so that their contribution at some location x_1 to the total intensity dI is reduced to

$$dI = j\rho(x_1)e^{-\tau(x_1)}dx_1 \tag{1.21}$$

where j is the *emissivity*. Integrating now gives us

$$I = \int j\rho(x_1)e^{-\tau(x_1)}dx_1 = \int_0^x S(\tau_1)d\tau_1 \tag{1.22}$$

Here τ is the total (optical) thickness of the slab, S is the so-called source function ($S = j/k$), and we assume that the intensity at the base of the slab is zero. In differential form, we can write the *equation of radiative transfer,*

$$\frac{dI}{d\tau} = I - S(\tau) \tag{1.23}$$

While we've taken I, τ, opacity κ, and emissivity j to be monochromatic, in reality all are general functions of wavelength and we write I_ν, τ_ν, κ_ν, and j_ν. κ_ν and j_ν in particular have both broad-band and narrow-band components. The dominant contributions to opacity in stellar interiors fall into these categories:

[10] Or sometimes $cm^2\ g^{-1}$, for *cgs* enthusiasts.

- Bound–Bound (bb): This represents the absorption of photons by atoms, leading to atomic transitions within the atom. They are thus responsible for the formation of spectral lines, such as the Lyman and Balmer series lines in hydrogen. Bound–bound transitions dominate opacity at low temperatures ($3.5 < \log T < 4.0$); at higher temperatures most atoms are fully ionized. The wavelength dependence of bound–bound opacity is complex and not easily approximated.
- Bound–Free (bf): Here atoms absorb a photon, leading to the ionization of an electron; also known as photoionization. Bound–free transitions are typically important at somewhat higher temperatures than bound–bound, $3.8 < \log T < 4.2$. Bound–free absorption can be reasonably well approximated by a law in which $\kappa_{\rm bf} \sim \rho T^{-3.5}$, a so-called "Kramer's opacity law."
- Free–Free (ff): Free–free transitions, also known as *bremsstrahlung*, involve the absorption of a photon by an electron, in the presence of a nearby nucleus (for momentum conservation reasons). Free–free processes become important at higher temperatures, and usually dominate for $\log T > 5.0$ when the plasma is fully ionized. Like bound–free absorption, free–free opacity can be reasonably well approximated by a Kramers-style opacity law in which $\kappa_{\rm ff} \sim \rho T^{-3.5}$.
- Electron Scattering (es): This is Thomson scattering from free electrons, and is a true scattering process rather than absorption, so represents photons simply changing direction rather than being rapidly absorbed and then re-emitted. It also (for nonrelativistic electrons) is "gray" in the sense that its value is wavelength-independent. Electron scattering is important at very high temperatures, $\log T > 6.0$.

This is an incomplete list, and among other things doesn't address molecular absorption or H^- absorption, both of which become significant only in the atmospheres of cooler stars and can be (generally) neglected in stellar interiors. Figure 1.6 shows the relative importances of the main contributions to total opacity[11] as a function of radius in the Sun.

1.5 Convective Transport

In *thermodynamic equilibrium* it can be shown (see, e.g., Mihalas 1978) that the source function reduces to the Planck function which describes a blackbody,

$$B_\nu(T) = \frac{2h\nu^3}{c^2} \frac{1}{e^{h\nu/kT} - 1} \tag{1.24}$$

Of course, the fact that a temperature gradient exists within a stellar interior implies that the interior is *not* in thermodynamic equilibrium, but the small scale of the gradient in all but near-surface layers means that *local thermodynamic equilibrium*

[11] Really to the Rosseland mean opacity; see below.

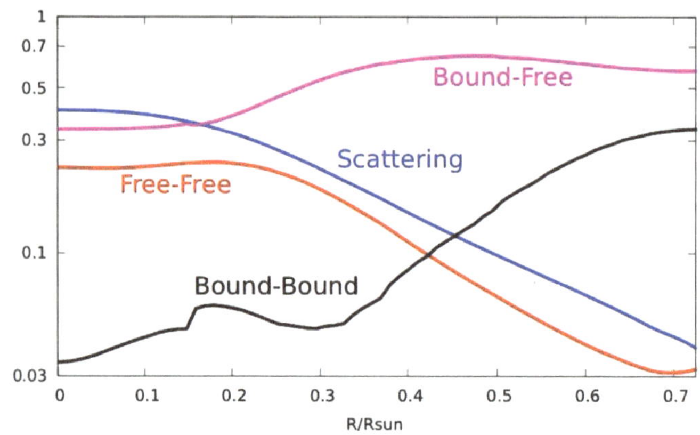

Figure 1.6. Relative contributions of different types of opacities to the Rosseland mean opacity as a function of solar radius. Reprinted with permission from Krief et al. (2016).

(LTE) is a reasonable approximation deep inside a star; in that case the Planck function is an excellent approximation to the local source function.

When the Planck function applies, we can make use of a number of simplifications. One of the most useful of these is that the energy density u for blackbody radiation is proportional to the fourth power of temperature, so

$$u = \frac{1}{3}aT^4 \tag{1.25}$$

where a is the radiation density constant. Now consider a star divided into onion-like layers. The temperature (and thus radiation flux and energy density as well) will fall as we move outwards from the center, and we can write the radiation flux at some radius r, or F_r, as

$$F_r = -\frac{d}{dr}\left(\frac{aT^4}{3}\right) \times c \times mfp \tag{1.26}$$

Here the first term represents the gradient of the energy density, the second term is the rate at which energy flows (the speed of light c), and the third is the mean free path for a typical photon. From the definition of opacity, we can write the mean free path λ as

$$\lambda = \frac{1}{\kappa\rho}, \tag{1.27}$$

while the negative sign ensures that the flow of energy is from the higher temperature regions at smaller radius toward the surface. Taking the derivative and making substitutions gives

$$F_r = -\frac{4ac}{3\kappa\rho}T^3\frac{dT}{dr} \tag{1.28}$$

We are really interested in the temperature gradient dT/dr, so rearranging gives

$$\frac{dT}{dr} = -\frac{3\kappa\rho}{4ac}\frac{F_r}{T^3} = -\frac{3\kappa\rho}{16\pi acr^2}\frac{L_r}{T^3} \tag{1.29}$$

where in the last step we've substituted for the luminosity at radius r,

$$F_r = \frac{L_r}{4\pi r^2} \tag{1.30}$$

Examination of this result shows that the radiative temperature gradient dT/dr increases when the opacity increases, which makes sense since higher opacity makes energy transport more difficult.

One still-open question here is what opacity κ to use. As noted earlier, opacity in general is a function of wavelength, but the temperature gradient can't be a function of wavelength! This implies the correct choice of opacity must be some kind of average opacity, and in fact we generally use the *Rosseland mean opacity*, which is

$$\frac{1}{\kappa} = \frac{\int_0^\infty B_\nu \kappa_\nu^{-1}}{B_\nu}, \tag{1.31}$$

where B_ν is the *Planck function*,

$$B_\nu(T) = \frac{2h\nu^3}{c^2}\frac{1}{e^{h\nu/kT} - 1} \tag{1.32}$$

which describes blackbody radiation. Physically speaking, the Rosseland mean opacity more heavily weights contributions from frequencies (or wavelengths) where the gas interacts most with radiation.

Radiation is not the only possible mechanism for energy transport within a star. Consider a parcel of gas at equilibrium within the star and displace it in the direction of decreasing pressure. If we allow the parcel to expand adiabatically so that it returns to pressure equilibrium in its new position, we can test for stability in the usual way, by asking the direction of the net force on the parcel. If that force is a restoring force, then the equilibrium of the parcel is stable; otherwise it is unstable to *convection*, which is the mass motion of material and an energy transport mechanism.

Consider a parcel of gas with some pressure P, volume V, and internal energy U. Dimensionally, pressure is energy density, so we can write the internal energy of a simple (monoatomic) gas as

$$PV = (\gamma - 1)\, U \tag{1.33}$$

where $(\gamma - 1)$ represents a cleverly-chosen proportionality constant. We can differentiate to write the change in the internal energy as

$$dU = \frac{1}{\gamma - 1}(PdV + VdP) \tag{1.34}$$

If the parcel is *adiabatic* so there are no energy flows across the boundary of the parcel, then we can also write

$$PdV = -dU \tag{1.35}$$

because all the PdV work we do on the parcel goes into changing the internal energy of the gas. We now have two expressions for dU, and they must be equal, so we can write

$$\frac{1}{\gamma - 1}(PdV + VdP) = -PdV \tag{1.36}$$

Rearranging gives

$$-\gamma PdV = VdP \tag{1.37}$$

which we can solve to get

$$P \propto V^{-\gamma} \tag{1.38}$$

or, equivalently,

$$P \propto \rho^{\gamma} \tag{1.39}$$

where γ is the *adiabatic exponent*. Usefully, we can also write it as

$$\gamma = \frac{d \log P}{d \log \rho}. \tag{1.40}$$

For an ideal monoatomic gas, $\gamma = 5/3$, but for more complex situations, such as a partially ionized gas, $\gamma < 5/3$ because the number of free particles is not constant and because internal energy can go into changing the ionization state. In such cases, we can make use of a generalized ionization exponent γ.

1.5.1 Adiabatic Gradient

Let's consider a parcel of gas at equilibrium within a star and displace it in the direction of decreasing pressure, and examine how the temperature of the parcel changes as this occurs. For simplicity, we consider an ideal gas with uniform composition and no changes in ionization during the process. In that case we start with the ideal gas law

$$P = \rho kT \tag{1.41}$$

During the radial excursion, the pressure of the parcel changes by some dP, which we can write as

$$dP = k\left(\rho\frac{dT}{dr} + T\frac{d\rho}{dr}\right)dr = \left(\frac{P}{T}\frac{dT}{dr} + \frac{P}{\rho}\frac{d\rho}{dr}\right)dr \tag{1.42}$$

where in the second step we make use of the original ideal gas law. If the parcel is adiabatic, then we know that it obeys a power law of the form

$$P \sim \rho^{\gamma} \tag{1.43}$$

and we can use this to write a different expression for dP

$$dP = \gamma\rho^{\gamma-1}\frac{d\rho}{dr}dr = \gamma\frac{P}{\rho}\frac{d\rho}{dr}dr \tag{1.44}$$

Once again, we've used the original power law to simplify in the last step. If the two expressions are both correct, then we can equate them to get

$$\gamma\frac{P}{\rho} = \frac{P}{T}\frac{dT}{dr} + \frac{P}{\rho}\frac{d\rho}{dr} \tag{1.45}$$

Rearranging gives

$$\frac{dT}{dr} = (\gamma - 1)\frac{T}{P}\frac{P}{\rho}\frac{d\rho}{dr} \tag{1.46}$$

and we can simplify further by substituting $P \sim \rho^{\gamma}$ to get

$$\left(\frac{dT}{dr}\right)_{\mathrm{ad}} = \frac{\gamma - 1}{\gamma}\frac{T}{P}\frac{dP}{dr} \tag{1.47}$$

This is the adiabatic gradient for temperature, sometimes written in terms of a logarithmic derivative

$$\frac{d\log T}{d\log P} = \frac{\gamma - 1}{\gamma} \tag{1.48}$$

Intuitively, you can think of the adiabatic gradient as the background temperature gradient, which is present in the absence of anything more interesting going on, like convective energy transport. But what happens in such a more interesting case? If we denote perturbed quantities by a superscript asterisk we have the situation shown in Figure 1.7. At the start, before we move the parcel, both pressure and density have the same values both inside the parcel and in the environment, so $\rho_1^* = \rho_1$ and $P_1^* = P_1$, where we use the 1 subscript to denote the initial position. After the parcel has moved and is again in pressure equilibrium, we again have $P_2^* = P_2$, but the density will in general be different. If the slope of the relation $dP/d\rho$ is steeper than the adiabatic relationship, then $\rho^* < \rho_2$. Then the buoyant force will be upwards, the parcel will keep rising, and the system is unstable—the criterion for convection. If $dP/d\rho$ is shallower than the adiabatic relationship, we will have stability. This allows us to write the stability criterion as

$$\left(\frac{dP}{d\rho}\right) < \left(\frac{dP}{d\rho}\right)_{\mathrm{ad}} \tag{1.49}$$

Let's rewrite this in a more useful form, in terms of the temperature gradient. Divide Equation (1.30) by Equation (1.29) to get

Figure 1.7. A cartoon illustration of the conditions for convection. If a parcel of gas with ρ_1, P_1 the same as the surrounding medium is vertically displaced to an environment with ρ_1, P_1, the condition for instability (convection) is that the new density ρ_* within the parcel is less than that of the new surrounding medium ρ_2.

$$\frac{dP}{P} = \frac{d\rho}{\rho} + \frac{dT}{T} \tag{1.50}$$

Dividing through by the left hand side makes this

$$1 = \frac{P}{\rho}\frac{d\rho}{dP} + \frac{P}{T}\frac{dT}{dP} \tag{1.51}$$

which solves to give

$$\frac{\rho}{P}\frac{dP}{d\rho} = \left[1 - \left(\frac{P}{T}\frac{dT}{dP}\right)_{\text{ad}}\right]^{-1} \tag{1.52}$$

We can use this result in the stability criterion above, Equation (1.49), to make it

$$\left[1 - \left(\frac{P}{T}\frac{dT}{dP}\right)\right]^{-1} = \left[1 - \left(\frac{P}{T}\frac{dT}{dP}\right)_{\text{ad}}\right]^{-1} \tag{1.53}$$

Admittedly this looks a bit confusing! However, we know that both sides of the original inequality are positive (because pressure increases linearly with density), so we can work through the algebra to get

$$\left(\frac{P}{T}\frac{dT}{dP}\right) < \left(\frac{P}{T}\frac{dT}{dP}\right)_{\text{ad}} \tag{1.54}$$

The usual notation is to use ∇ to represent the logarithmic derivative, so we can write the stability criterion as

$$\nabla > \nabla_{\text{ad}} = \frac{\gamma - 1}{\gamma} \tag{1.55}$$

which is the Schwarzschild stability criterion. Note that in this discussion we have assumed that chemical composition is constant. This may not be true! For example, in the radiative core of a star evolved away from the ZAMS there is a compositional gradient due to the increased nuclear reaction rate nearer the core. In the stellar envelope, changes in ionization state can also change the mean molecular weight and thus the chemical composition. A more general version of the stability criterion takes that into account, so

$$\nabla > \nabla_{\text{ad}} + \nabla_{\mu}\,, \tag{1.56}$$

the Ledoux criterion.

If the stability criterion is **not** satisfied, then in general we will have convective motions, with moving parcels of gas acting to reduce the temperature gradient to the point where the stability criterion is only just satisfied. There is a complication, however! Our simple argument suggests that parcels of gas remain coherent entities indefinitely, which would imply that rising parcels also continue to accelerate indefinitely. This would imply extremely large energy transport by convection, and of course is also in contrast to what we actually observe in the Sun. Instead, we assume that the parcel remains a coherent entity for some distance comparable, to within a factor of a few, to the atmospheric pressure scale height,

$$H_{\text{P}} = \frac{kT}{\mu m_{\text{H}} g} \tag{1.57}$$

This distance is known as the *mixing length* l_{ML}, and it is parameterized as

$$l_{\text{ML}} = \alpha_{\text{ML}} H_{\text{P}} \tag{1.58}$$

where H_{P} is the pressure scale height.

At that point, we would expect the temperature gradient to be only *slightly* more than the adiabatic gradient. How much more? A formal comparison gives the result that convective transport inside stars is so efficient that in most places once convection occurs it forces the actual temperature gradient to be equal to the adiabatic gradient to within some factor δ, which we can comfortably take to be negligible.

Where inside a star is convection likely to be important? We can identify those regions by revisiting the radiative temperature gradient

$$\frac{dT}{dr} = -\frac{3\kappa\rho}{16\pi a c r^2}\frac{L_r}{T^3} \sim \kappa L_r\left(\frac{\rho}{T^3}\right) \tag{1.59}$$

Most obviously, this becomes large when the opacity κ is large and radiative transport is clearly impeded. Broadly speaking this occurs at lower temperatures, so we would expect to see convective transport becoming important in the outer layers of lower-mass stars on the main sequence. The radiative temperature gradient is also large when L_r is large, which occurs where energy generation is highly concentrated. On the main sequence, this is near the cores of higher-mass stars which use the CNO cycle as their dominant energy production mechanism, so we would anticipate seeing convective transport becoming important near the cores of these stars. Finally, convective transport will dominate when ρ/T^3 is large, which also occurs in the outer layers of lower main sequence stars.

1.6 Calculating Stellar Models

Above we've derived the equations of stellar structure,

$$\begin{aligned} \frac{dM_r}{dr} &= 4\pi r^2\rho \\ \frac{dL_r}{dr} &= 4\pi r^2\rho\varepsilon \\ \frac{dP}{dr} &= -\frac{GM_r\rho}{r^2} \\ \left(\frac{dT}{dr}\right)_{\mathrm{rad}} &= -\frac{3\kappa\rho}{16\pi acr^2}\frac{L_r}{T^3} \\ \left(\frac{dT}{dr}\right)_{\mathrm{conv}} &= \frac{\gamma-1}{\gamma}\frac{T}{P}\frac{dP}{dr} \end{aligned} \tag{1.60}$$

The opacity as a function of wavelength, or a mean opacity, is required in order to complete the radiative temperature gradient expression, and this can be obtained from calculated tables of opacity (such as Iglesias & Rogers 1996) or from approximations if high precision is not required. Furthermore, we have the thermodynamic variables P, T, and ρ, and these are coupled to one another through an *equation of state* (EOS). The simplest EOS is just the ideal gas law,

$$P = \frac{\rho kT}{\mu_{\mathrm{I}} m_{\mathrm{H}}} \tag{1.61}$$

Deeper in the stellar interior, the plasma is essentially fully ionized, so we need to account for electron pressure as well as ion pressure, and we can write

$$P = P_{\mathrm{e}} + P_{\mathrm{I}} = \frac{\rho kT}{\mu_{\mathrm{e}} m_{\mathrm{H}}} + \frac{\rho kT}{\mu_{\mathrm{I}} m_{\mathrm{H}}} \tag{1.62}$$

where the mean molecular weights depend on the composition and ionization state of the plasma. In upper main sequence stars, radiation pressure becomes important and must also be included, while in lower-mass stars and giants matter in the core becomes partially degenerate, adding another term to consider. In white dwarf interiors we also need to consider Coulomb interactions and phase transitions associated with crystallization.

The stellar structure equations are one-dimensional, depending only on radius r, which reflects the spherical symmetry of the problem.[12] To solve for the internal structure, we solve these coupled equations given constraints: mass M, initial composition X, Y, and Z, and mixing length parameter $\alpha_{\rm ML}$ (alternatively, radius can be fixed and mixing length allowed to vary). Of course, as with any set of differential equations, solution requires boundary conditions. These are straightforward at the center ($M_0 = 0$, $L_0 = 0$), but somewhat more complex at the surface, though we can fix the luminosity at $L = 4\pi R_*^2 \sigma T_e^4$ and the pressure $P(r)$ through a simple stellar atmosphere model. In computational practice, derivatives are normally rewritten in terms of dm rather than dr, so dP/dr becomes

$$\frac{dP}{dm} = \frac{dP}{dr}\frac{dr}{dm} = -\frac{GM_r\rho}{r^2}\frac{1}{4\pi r^2\rho} = -\frac{GM_r}{4\pi r^4} \tag{1.63}$$

and similarly for the other structure equations. Departures from spherical symmetry due to, e.g., rotation, are typically either ignored (if small enough) or handled as perturbations from spherical solutions (if larger). Static spherically symmetric models based on these stellar structure equations necessarily ignore the presence of rotation, magnetic fields, chemical diffusion and settling, convective overshoot, mass loss, meridional circulation, and binary star interactions, among other effects. Fully detailed inclusion of these is still in early stages, and all are subjects of ongoing research.

One important use of stellar models is to explore stellar *evolution*: changes in composition, radius, and internal structure over time. We do this by adjusting the composition $[X(r),\ Y(r),\ Z(r)]$ based on calculated nuclear reaction rates over some timestep Δt, recalculating the model for some later time $t = t_0 + \Delta t$. While the models described above are static, significant changes in stellar composition occur on the nuclear timescale, and other structural changes on the thermal timescale. Both of these are long enough that the timestep Δt is generally not unreasonably long, and its length can be dynamically adjusted to align with the scale of ongoing changes, in order to afford sufficient time resolution during more rapid changes. However, the most rapid events, such as the helium flash and supernova explosions, must still be modeled using specialized or ad hoc techniques.

While there is general agreement on the equations to be solved, there is somewhat more diversity in the computational techniques used to solve them, particularly as some codes are optimized for different types of stars and/or different stages of their evolution. Table 1.1 shows a variety of codes in common use for calculating stellar interiors and evolution modes.

[12] This is a simplification which neglects the influence of non-spherically-symmetric processes such as rotation and magnetic fields.

Table 1.1. Summary of Some Stellar Structure and Evolution Codes

Name	Acronym	Nuclear Reactions	Low-T Opacity	High-T Opacity	Equation of State	Diffusion	Mass Loss
Aarhus Stellar Evolution Code (Christensen-Dalsgaard, 2008)	ASTEC	Clayton (1984) (w/ mods)	Ferguson et al. (2005)	OPAL (Rogers & Iglesias, 1992)	OPAL EOS (Rogers & Nayfonov, 2002)		
Bag of Stellar Tricks and Isochrones (Pietrinferni et al. 2007)	BaSTI	Angulo et al. 1999	Ferguson et al. (2005)	OPAL	FreeEOS (Irwin, 2012)	Schlattl & Salaris (2003)	Reimers
Code d'Evolution Stellaire Adaptatif et Modulaire	CESAM2K (Morel & Lebreton, 2008)	NACRE (Xu et al. 2013)	Alexander & Ferguson (1994)	OPAL	multiple options	Michaud & Proffitt (1993)	
GARching STellar Evolution Code (Weiss & Schlattl, 2008)	GARSTEC	NACRE	Ferguson et al. (2005)	OPAL	OPAL EOS	Thoul et al. (1994)	Reimers
La Plata stellar evolution Code (Althaus et al. 2002)	LPCODE	Caughlan & Fowler (1988)	Ferguson et al. (2005)	OPAL	custom	Iben & MacDonald (1985)	*ad hoc*
Modules for Experiments in Stellar Astrophysics Paxton et al. (2015)	MESA	Caughlan & Fowler (1988) + NACRE	Ferguson et al. (2005)	OPAL	OPAL EOS	Thoul et al. (1994)	various
Monash version of the Mt. Stromlo evolution code (Constantino et al. 2014)	MONSTAR	JINA REACLIB (Cyburt et al. 2010)	Marigo & Aringer (2009)	Iglesias & Rogers (1996)	Timmes & Arnett (1999)		
Yale Rotation Stellar Evolution Code (Demarque et al. 2008)	YREC	NACRE or custom	Ferguson et al. (2005)	OPAL	OPAL EOS	Thoul et al. (1994)	

Beyond ease of use, perhaps the greatest variation between codes is in the input physics, particularly the choices made for initial elemental abundances and opacities, for nuclear reaction rates, for convection, and for the specific choice of surface boundary condition (typically a $T - \tau$ relation or simple or tabulated model atmosphere). However, other issues to consider include treatment of diffusive settling and radiative levitation of heavier elements, mass-loss, rotation and its effects on mixing and convection, convection itself and how and whether convective overshoot at the convective–radiative boundary is included. Some implementations provide choices of how these kinds of processes are included, while others allow the user to add or modify details of their treatment, while still others make the choice for the user. Beyond this, more specialized codes can treat turbulence, magnetic fields, MHD effects, energy transport by gravity waves, stellar winds, and the interaction between pulsation and convection. Inclusion of many of these higher-order effects can require two-dimensional or fully three-dimensional implementations, with substantial impact on computational cost, which can sometimes be finessed by parameterizing the impacts seen in limited multi-dimensional runs and trying to include them in one-dimensional codes.

If complete interiors models are not required, then grids of various model runs are also available. These typically span a wide range of input metallicities, initial helium abundances, and ages, allowing (and in most cases providing) isochrones as well as evolutionary tracks. For ease of comparison with observations, the tracks and isochrones are generally transformed to observables in a wide range of different photometric systems. Sources include DSEP (Dartmouth Stellar Evolution Program, Dotter et al. 2008), YaPSI (Yale–Potsdam Stellar Isochrones, Spada et al. 2017), PARSEC (Padova Trieste Stellar Evolution Code, Bressan et al. 2012), and MIST (MESA Isochrones & Stellar Tracks, Dotter 2016; Choi et al. 2016; Paxton et al. 2015), and others.

1.7 Stellar Evolution

In general, and encouragingly, results from the different codes agree in a broad sense, giving us confidence that we understand stellar evolution at a fairly fine-grained level. Stars begin their main sequence lives with a structure determined by their mass: very low-mass stars (below $\sim 0.4 M_\odot$, depending on compositional details) remain fully convective throughout, more massive stars like the Sun use the *p–p* chain and have radiative cores and convective envelopes, while the most massive stars (above $\sim 1.2 M_\odot$, again a function of metallicity Z) make use of the CNO cycle and develop convective cores and radiative envelopes. Once hydrogen fusion becomes the main energy source for the star, they are on the Zero-Age Main Sequence (ZAMS).

Core hydrogen fraction falls with time, as hydrogen fusion continues. In stars with convective cores, mixing means that the entire core depletes hydrogen at the same rate, while in stars with radiative cores, composition changes radially. In either case, core hydrogen is eventually exhausted, and the star reaches the Terminal-Age Main Sequence (TAMS). At this point, hydrogen fusion continues in a shell around the core, leading to contraction of an inert (non-energy-producing) helium-rich core and expansion of the overlying hydrogen envelope. On the H–R diagram, the star becomes first a subgiant and then a red giant.

On the red giant branch, the star expands significantly and undergoes a massive increase in luminosity. The deepening convective envelope at this stage can reach down to parts of the star that have previously experienced fusion, convecting them up to the surface in a process known as "dredge-up," which can be visible spectroscopically. The high luminosity at this stage means that the stellar residence time on the red giant branch is typically $<10\%$ of the main sequence lifetime.

As the helium core grows in mass, it continues to compress and its temperature increases. From here on, further evolution is highly sensitive to mass. The He cores of the lowest-mass stars never reach the $\sim 10^8$ K temperature required to activate the triple-alpha process, and the star will eventually run out of nuclear fuel in the shell and the core/shell remnant will contract to become a white dwarf. Higher-mass stars transition to helium fusion in the core (via a helium flash for intermediate-mass stars and more sedately for higher-mass ones), while eventually hydrogen shell fusion ceases. At this point the star in on the horizontal branch, where it remains until an inert C/O core develops, in parallel with the earlier development of an inert He core.

Further evolution leads to renewed envelope expansion and an ascent up the asymptotic giant branch (AGB) on the HR diagram, with additional dredge-up of He fusion products. As the star continues to ascend the AGB, hydrogen shell fusion restarts farther out, potentially leading to thermal pulsing drive by He layer fusion events, and yet another dredge-up of CNO products and s-process elements built by the pulses. At the top of the AGB, outflows intensify into a superwind, which strips away much of the envelope to form a planetary nebula, while fusion ceases and the remaining core and outer envelope become a white dwarf.

As their cores contract, more massive stars become hot enough to fuse elements beyond He in layers, creating an onion structure in which mean molecular weight μ increases in steps moving inwards toward the core, with alternating layers undergoing fusion and remaining inert. These stars have Eddington-limited luminosities, where $L \sim M$, so luminosity is only weakly correlated with stellar mass. The dense, extremely hot ($T \sim 10^9$K), compact stellar core is primarily supported by electron degeneracy pressure, but eventually when Fe fusion begins in the core, loss of energy from that process allows gravity to overwhelm that pressure source, leading to a core-collapse supernova and a dramatic end to the life of the star.

Figure 1.8 shows the evolutionary tracks for stars of various masses on an H–R diagram on the left, while on the right are the corresponding *isochrones*, lines of constant age. The tracks show the evolution of intermediate-mass stars into white dwarfs, while the isochrones show that the lifetimes of higher-mass stars are far shorter than those of lower mass.

1.8 Model Comparisons

Given the differences in input physics and other implementation details, it's valuable to consider how similar results are, when produced by different codes using the same

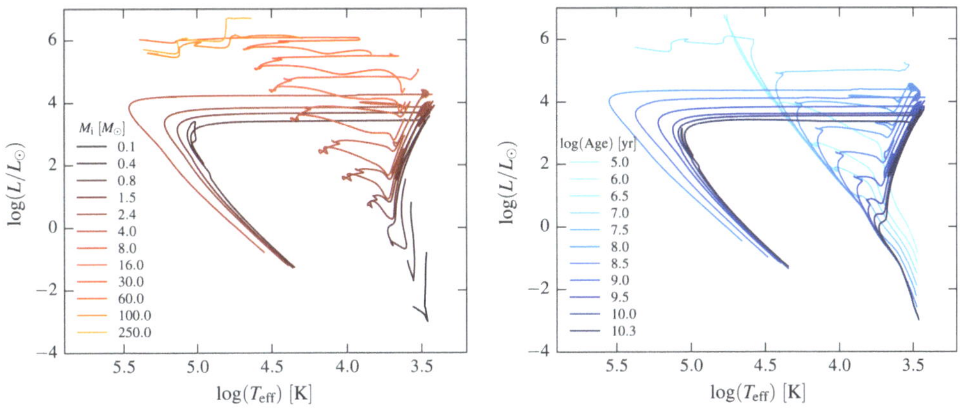

Figure 1.8. An example solar-metallicity grid of stellar evolutionary tracks (left) and isochrones (right) covering a wide range of stellar masses, ages, and evolutionary phases. Reprinted with permission from Choi et al. (2016).

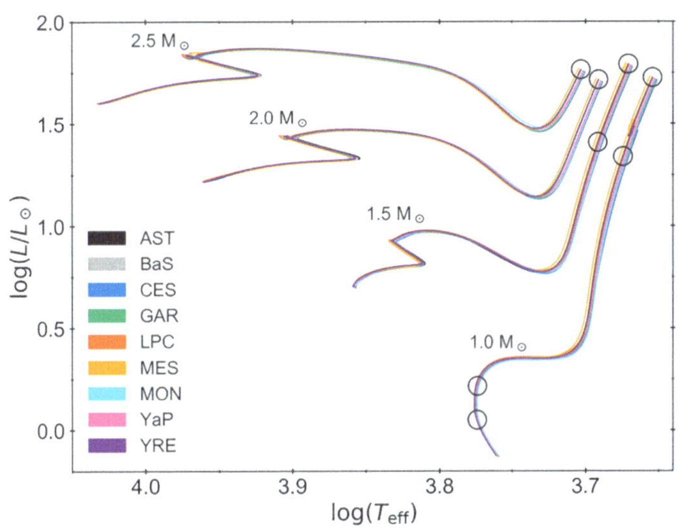

Figure 1.9. Hertzsprung–Russell diagram (HRD) of solar-radius calibrated science cases for all codes evaluated by Silva Aguirre et al. (2020a). Evolutionary tracks are shown, and the open circles indicate the positions of stellar models subjected to detailed comparison. Reprinted with permission from Silva Aguirre et al. (2020a).

input parameters. We are not aiming for an exhaustive comparison here, simply an illustrative one.

Silva Aguirre et al. (2020a, 2020b) performed a detailed comparison of nine different stellar evolution codes, comparing results on the red giant branch for stars ranging from $1 - 2.5 M_{\odot}$, with results shown in Figure 1.9. All codes were tuned to reproduce the current solar observables at an age of 4.57 Gyr by adjusting the mixing length parameter $\alpha_{\rm ML}$ (typically between 0.5 and 2.5; Joyce & Tayar 2023) at

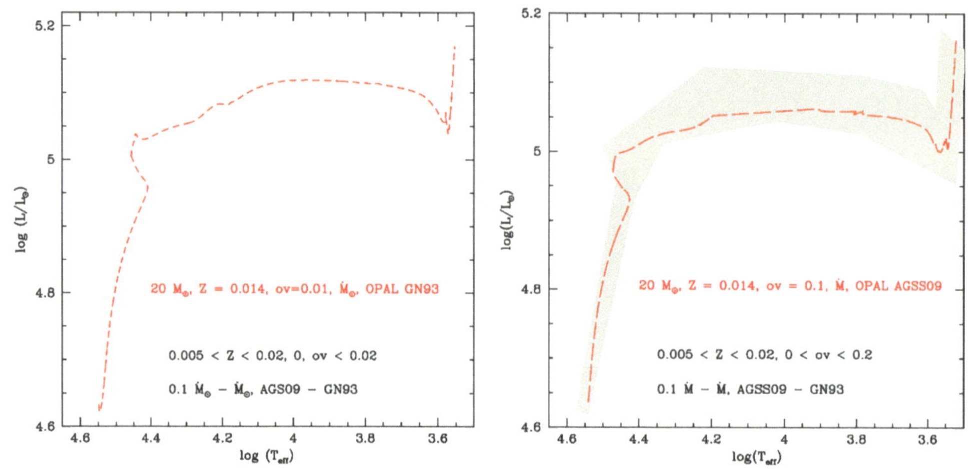

Figure 1.10. The region occupied on the HR diagram by the evolutionary tracks of $20M_\odot$ models computed with MESA (left) and STAREVOL (right) with different opacities, metallicities, mass loss rates and overshooting parameters. The dashed red line shows the track of the standard model, with the parameters as indicated on the figure (the overshooting parameter is not defined in the same way in both codes, hence the different values). The shaded envelope defines a rough global intrinsic uncertainty on $20M_\odot$ models. Reprinted with permission from Martins & Palacios (2013).

a fixed initial abundance of $X = 0.70$, $Y = 0.28285$, $Z = 0.01715$. When on the main sequence, identical inputs then led to a spread in effective temperature of roughly ~20 K, independent of stellar mass, and a range in relative convective core mass M_C/M_* of ~0.01.

On the RGB, the effective spread in effective temperature doubled to ~40 K, while age differences over the sample examined showed variations of 2%–5%, monotonically increasing with stellar mass. These results are consistent with unpublished work by Monteiro and collaborators (https://www.astro.up.pt/corot/welcome/meetings/m5/ESTA_CW9_Monteiro_3.pdf), in which they compare results for seven different stellar evolution codes and find spreads as large as 1.4% in radius, 1.7% in effective temperature, 8.3% in luminosity, and 7.2% in age for a $0.9M_\odot$ star on the main sequence, and roughly comparable model-based uncertainties for subgiants.

For more massive stars, Martins & Palacios (2013) found "large differences in luminosity and temperature," as visible in Figure 1.10, implying that model uncertainties here are considerably larger than farther down the HR diagram.

References

Alexander, D. R., & Ferguson, J. W. 1994, ApJ, 437, 879

Althaus, L. G., Serenelli, A. M., Córsico, A. H., & Benvenuto, O. G. 2002, Mon. Not. R. Astron. Soc., 330, 685

Andrae, R., Fouesneau, M., Creevey, O., et al. 2018, A&A, 616, A8

Angulo, C., Arnould, M., Rayet, M., et al. 1999, NuPhA, 656, 3

Baines, E. K., Clark, J. H., Schmitt, H. R., Stone, J. M., & von Braun, K. 2023, AJ, 166, 268
Borucki, W. J., Koch, D., Basri, G., et al. 2010, Sci, 327, 977
Bressan, A., Marigo, P., Girardi, L., et al. 2012, MNRAS, 427, 127
Casagrande, L., & VandenBerg, D. A. 2018, MNRAS, 479, L102
Caughlan, G. R., & Fowler, W. A. 1988, ADNDT, 40, 283
Chandrasekhar, S. 1939, An Introduction to the Study of Stellar Structure (Chicago, IL: Univ. Chicago Press)
Choi, J., Dotter, A., Conroy, C., et al. 2016, ApJ, 823, 102
Christensen-Dalsgaard, J. 2008, Astrophys. Space Sci., 316, 13
Clayton, D. D. 1984, Principles of stellar evolution and nucleosynthesis (Chicago, IL: Univ. Chicago Press)
Constantino, T., Campbell, S., Gil-Pons, P., & Lattanzio, J. 2014, Astrophys. J., 784, 56
Cyburt, R. H., Amthor, A. M., Ferguson, R., et al. 2010, ApJS, 189, 240
Demarque, P., Guenther, D. B., Li, L. H., Mazumbar, A., & Straka, C. W. 2008, Astrophys. Space Sci., 316, 31
Dotter, A. 2016, ApJS, 222, 8
Dotter, A., Chaboyer, B., Jevremović, D., et al. 2008, ApJS, 178, 89
Eddington, A. S. 1926, The Internal Constitution of the Stars (Cambridge: Cambridge Univ. Press)
Emden, R. 1927, NW, 15, 769
Ferguson, J. W., Alexander, D. R., Allard, F., et al. 2005, ApJ, 623, 585
Fetherolf, T., Welsh, W. F., Orosz, J. A., et al. 2019, AJ, 158, 198
Gaia CollaborationVallenari, A., Brown, A. G. A., et al. 2023, A&A, 674, A1
Gray, D. F., & Kaur, T. 2019, ApJ, 882, 148
Iben, I. Jr., & MacDonald, J. 1985, Astrophys. J*, 296, 540*
Iglesias, C. A., & Rogers, F. J. 1996, ApJ, 464, 943
Irwin, A. W. 2012, FreeEOS: Equation of State for stellar interiors calculations, Astrophysics Source Code Library, record ascl:1211.002
Joyce, M., & Tayar, J. 2023, Galaxies, 11,
Krief, M., Feigel, A., & Gazit, D. 2016, ApJ, 821, 45
Lindegren, L., Klioner, S. A., & Hernández, J. 2021, A&A, 649, A2
Marigo, P., & Aringer, B. 2009, A&A, 508, 1539
Martins, F., & Palacios, A. 2013, A&A, 560, A16
Michaud, G., & Proffitt, C. R. 1993, Astronomical Society of the Pacific Conference Series, Vol. 40, IAU Colloq. 137: Inside the Stars, ed. W. W. Weiss, & A. Baglin, 246
Mihalas, D. 1978, Stellar Atmospheres (San Francisco, CA: Freeman)
Morel, P., & Lebreton, Y. 2008, Astrophys. Space Sci., 316, 61
Paxton, B., Marchant, P., Schwab, J., et al. 2015, ApJS, 220, 15
Pietrinferni, A., Cassisi, S., Salaris, M., Cordier, D., & Castelli, F. 2007, in IAU Symposium Vol. 241, Stellar Populations as Building Blocks of Galaxies, ed. A. Vazdekis, & R. Peletier, 39–40
Pecaut, M. J., & Mamajek, E. E. 2013, ApJS, 208, 9
Ramsey, L. W., Adams, M. T., Barnes, T. G., et al. 1998, Proc. SPIE, 3352, 34
Rogers, F. J., & Iglesias, C. A. 1992, ApJS, 79, 507
Rogers, F. J., & Nayfonov, A. 2002, ApJ, 576, 1064
Schlattl, H., & Salaris, M. 2003, A&A, 402, 29

Schwarzschild, M. 1958, Structure and Evolution of the Stars (Princeton, NJ: Princeton Univ. Press)
Silva Aguirre, V., Christensen-Dalsgaard, J., Cassisi, S., et al. 2020a, A&A, 635, A164
Silva Aguirre, V., Christensen-Dalsgaard, J., Cassisi, S., et al. 2020b, A&A, 635, A164
Spada, F., Demarque, P., Kim, Y. C., Boyajian, T. S., & Brewer, J. M. 2017, ApJ, 838, 161
Then, I., & MacDonald, J. 1985, ApJ, 296, 540
Thoul, A. A., Bahcall, J. N., & Loeb, A. 1994, ApJ, 421, 828
Timmes, F. X., & Arnett, D. 1999, Astrophys. J. Suppl. Ser., 125, 277
Weiss, A., & Schlattl, H. 2008, Astrophys. Space Sci., 316, 99
Xu, Y., Takahashi, K., Goriely, S., et al. 2013,918, 61

Derek L Buzasi

Chapter 2

An Observational Perspective

Perhaps the first formal observation of a periodically variable star was by Fabricius in 1597 (Hoffleit 1997), though the variability may have been known in antiquity: the star was Mira (or Ceti). By 1638 Johannes Holwerda had determined the period of variability to be about 11 months, and the study of stellar pulsations had truly begun.

Why do stars oscillate? The waves responsible for the variability are acoustic in nature, so since such waves can't propagate in the space surrounding the star, they can be considered as bounded acoustic cavities, akin to musical instruments. Any acoustic disturbance then produces waves, which travel throughout the stellar interior at the local sound speed. However, just as a musical instrument is resonant at only certain frequencies and not at others, depending on its size, shape, and construction details, so too a star is resonant at only specific frequencies.

In order to have a propagating wave, we need a restoring force acting on the gas. In a star, there are essentially two possible restoring forces, leading to two different types of acoustic waves. The first possibility is pressure, where the wave propagates through successive compression and rarefaction of the medium, so that the wave is longitudinal, just as a sound wave in air. The second possible restoring force is gravity, where the wave is transverse and the medium essentially sloshes back and forth perpendicular to the propagation direction, like water waves near the beach. Pressure waves can propagate in a uniform medium, while gravity waves require non-uniformity; in a star, that typically means density stratification. The usual terminology is that we call pressure waves *p-modes* and gravity waves *g-modes*.

Just as the spectrum of notes produced by a musical instrument allows us to infer information about the instrument itself (a note played on a bass sounds different from the same note played on a violin), the spectrum of notes produced by a star allows us to infer information about both its overall characteristics such as mass and radius, and its internal characteristics such as the depth of the convection zone, age, composition, and internal dynamics.

doi:10.1088/2514-3433/ae03a0ch2

Of course, we can't detect the waves directly because sound doesn't travel in space, but we can detect changes in overall brightness caused by periodic increases and decreases in the size and temperature of the star; this is what Fabricius unknowingly did back in 1597. Mira's pulsations are purely *radial* and the period long, so its luminosity changes are large (up to a factor of $\sim$1000); other stars show non-radial oscillations, and for some of these the periodic changes in luminosity are at the parts-per-million level. Alternatively, rather than changes in luminosity, using spectroscopy, we can also detect periodic variability in velocity as the stellar photosphere moves in response to the acoustic waves; here the detectable signals range from km s^{-1} down to cm s^{-1} depending on the star. More esoteric techniques also exist, including small changes in the average temperature of the star as manifested in temperature-sensitive absorption features in the spectrum, or even changes in polarimetry caused by the small shape changes associated with non-radial oscillations.

All stars likely oscillate, though some may do so at an undetectable level. Figure 2.1 shows the H–R diagram from the point of view of an asteroseismologist. The cross-hatching visible for each type of star indicates the dominance of pressure (p-modes, \\) or gravity modes (g-modes, //), while the color indicates the effective photospheric temperature. Overplotted are the zero-age main sequence (ZAMS) and a few representative evolutionary tracks, while the roughly parallel solid lines running toward the top right from the main sequence represent the classical instability strip, home of "classical" (by which we really mean large amplitude) pulsators such as Cepheids and RR Lyrae stars, as well as somewhat less widely known examples such as roAp, δ Scuti, and γ Dor stars. In general, oscillation amplitudes in both radial velocity and luminosity increase and oscillation frequencies decrease as we move up the diagram and surface gravity decreases. Table 2.1 shows a selected list of oscillating star types.

Of course, oscillation detections in large-amplitude classical pulsators like Cepheids were successful from an early date, aided by the fact that both photometric and radial velocity oscillation amplitudes in these stars are large. Attempts at detection of solar-like oscillations proved less successful, however. Even with the advent of early electronic detectors, ground-based photometry was limited by scintillation noise, so efforts focused on radial velocity detections. These in turn required the appearance of stable ground-based instruments, fed by optical fibers to remove them from the hostile thermal and vibrational environment of the observatory, and provided with stable velocity references such as iodine cells or hollow-cathode lamps.

These early attempts concentrated on stars anticipated to have relatively large velocity amplitudes, such as Procyon (α CMi), η Boo, and α Cen A. An early attempt was Brown et al. (1991), who detected excess power in the periodogram of Procyon, though they didn't actually resolve modes, and this discovery was confirmed by Martić et al. (1999). Kjeldsen et al. (1999) attempted to measure periodic equivalent width variations in α Cen A caused by temperature changes during the oscillation cycle, and were able to establish an upper limit approaching the solar level. However, the first real detection was probably in the B2 IV star β Hyi (Bedding et al. 2001) using the University College London Echelle Spectrograph (UCLES) at the coudé focus of the 3.9 m Anglo-Australian Telescope (AAT). Over

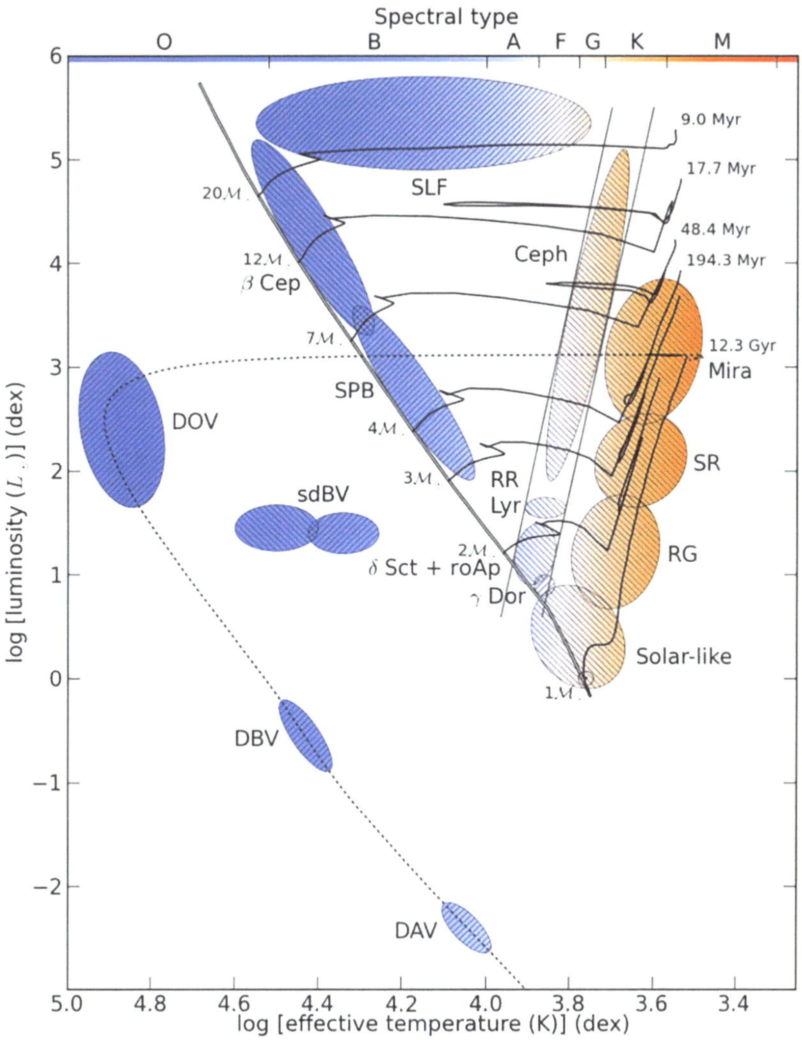

Figure 2.1. Luminosity-effective temperature (or Hertzsprung–Russell) diagram showing the positions of the main classes of pulsating variable stars, colored roughly according to effective temperature. The zero-age main sequence is are also shown, along with the classical instability strip, and evolutionary tracks for model stars of various masses. Hashes indicate the dominant visible pulsation mode type: p modes (\\) and g modes (//). Approximate spectral types are indicated on the top axis. Reprinted with permission from Pápics (2013).

5 nights and nearly 1200 measurements, they were able to clearly detect the presence of p-mode oscillations centered around $\nu_{\max} = 1.0\,\mathrm{mHz}$, with amplitudes peaking around $0.5\,\mathrm{m\,s^{-1}}$, as well as establishing a reasonable measurement of the large frequency separation.[1] Future ground-based efforts were aided by the construction of a new generation of fiber-fed echelle instruments intended primarily for exoplanet

[1] Both $\nu_{\max}$ and the large frequency separation $\Delta\nu$ are important asteroseismic diagnostics, as we shall see.

Table 2.1. An Incomplete List of Types of Oscillating Stars

Name	Typical Periods	Typical Luminosity Amplitude	Spectral Type	Notes
Mira variables	100–1000 d	>1 magnitude	M III	Thermally-pulsing AGB
δ Cephei stars	1–100 d	>1 magnitude	F–G Ib	Type I Cepheid (metal rich)
W Virginis stars	10–100 d	>1 magnitude	F–K Ib	Type II Cepheid (metal-poor)
RR Lyrae stars	2–24 h	0.2–1.2	A–F III	
δ Scuti stars	0.3–6 h	0.003–0.3	A–F V/IV	
β Cephei stars	3–7 h	0.01–0.3	B2–3 IV–V	p-mode pulsators
ZZ Ceti stars	30 s–20 m	0.001–0.2	A (DAV)	White dwarfs
GW Virginis stars	5–90 min	0.005–0.05	O (DOV)	Pre-white dwarfs
Rapidly oscillating Ap (roAp) stars	5–25 min	0.001–0.005	A	Also α^2~CVn variables
V777 Herculis stars	2–20 min	ppt	B (DBV)	White dwarfs
Slowly Pulsating B (SPB) stars	0.5–5 d	0.01–0.1	B V	g-mode pulsators
Solar-like oscillators	3–15 min	ppm	F–M V	p-mode pulsators
sdBVr	2–10 min	ppt	B	Extreme horizontal branch; also V361 Hydrae stars
γ Doradus stars	0.3–3 d	0.01–0.3	F–A V	g-mode pulsators
Solar-like giant oscillators	1–18 hr	ppt	G–M III	p-mode pulsators
sdBVs	1–3 hr	0.001–0.01	B	Extreme horizontal branch; also V1093 Herculis stars
sdOV	1–2 min	0.001–0.05	O (SDO)	Pulsating subdwarf O star
LBV	1 – 10 yr		B – K I	Also S Dor
α Cyg variables	Days–weeks	~0.1 mag	B–A	
SX Phe variables	Hours	~0.75 mag	A V	
R CrB variables	~months	~0.3 mag	F–G Ia	

detection, though the requirements for asteroseismology (particularly high-cadence, long baseline observations) only partially overlap the ideal instrumental configuration for exoplanets. For more discussion of the current state of the art, see Sections 4.3 and 4.4.

From early on, photometry from space was identified as having a key role in asteroseismology. Oscillations produce detectable stellar luminosity variations, and from space even a small aperture is capable of detecting this photometrically, due to the absence of a confounding terrestrial atmosphere. In addition, area detectors like CCDs hold out the promise of high efficiency, since they can monitor hundreds or thousands of stars simultaneously. The first real effort was the EVRIS experiment on MARS-96 (Buey et al. 1997), which planned an asteroseismology experiment en route to Mars; unfortunately, the spacecraft failed, but the idea was a sound one. The first success came from an unexpected source, when the Wide-Field Infrared Explorer (WIRE) satellite failed soon after launch, but the spacecraft retained a working star-tracker, which was converted for use as a science instrument (Buzasi 2002; Bruntt & Buzasi 2006). WIRE successfully detected p-mode oscillations in Procyon and Alpha Cen A, along with a number of red giants and classical pulsators (Schou & Buzasi 2001; Bruntt et al. 2005; Stello et al. 2008). The Hubble Space Telescope (HST) and the Solar Mass Ejection Imager (SMEI), the latter also working in a repurposed role, also contributed a few more detections (Edmonds & Gilliland 1996; Tarrant et al. 2007), but numbers truly began to climb when the Microvariability and Oscillations of STars (MOST) mission was launched in 2003 as the first purpose-built asteroseismology mission, returning exquisite light curves in stars ranging from solar-like to roAp stars, δ Scuti pulsators, red giants, and beyond (Walker et al. 2003; Matthews 1998, 2007). The CoRoT (Convection, Rotation, and Planetary Transits) satellite next came on line in 2006, detecting oscillations in a number of main sequence stars (Michel et al. 2008) and thousands of red giants (Hekker et al. 2009). The space revolution continued with Kepler (Chaplin et al. 2011), which grew the number of detections by another order of magnitude, and most recently TESS (Hatt et al. 2023), which extended Kepler's success to higher cadence and across nearly the entire sky. The HR diagram shown in Figure 2.2 demonstrates the impact of industrial-scale space photometry. Here the inset shows detections prior to the launch of CoRoT, while the main diagram shows detections due to Kepler. The reason for the bias toward detections of red giants is clear as well: their oscillations are of higher amplitude, making them easier to detect, and the stars are intrinsically bright, making them easier to observe. The revolution continues with smaller specialized telescopes, such as CHEOPS (Moya et al. 2018) and BRITE (Weiss et al. 2021), while TESS continues operations as of this writing.

2.1 Upper Main Sequence Stars

In a star, energy is generated in the deep interior and flows toward the surface where it is radiated into space; an excitation mechanism itself is a general term for a number of processes capable of intercepting that radiation and transforming a small portion of it into mechanical pulsational energy. The fraction of the available energy

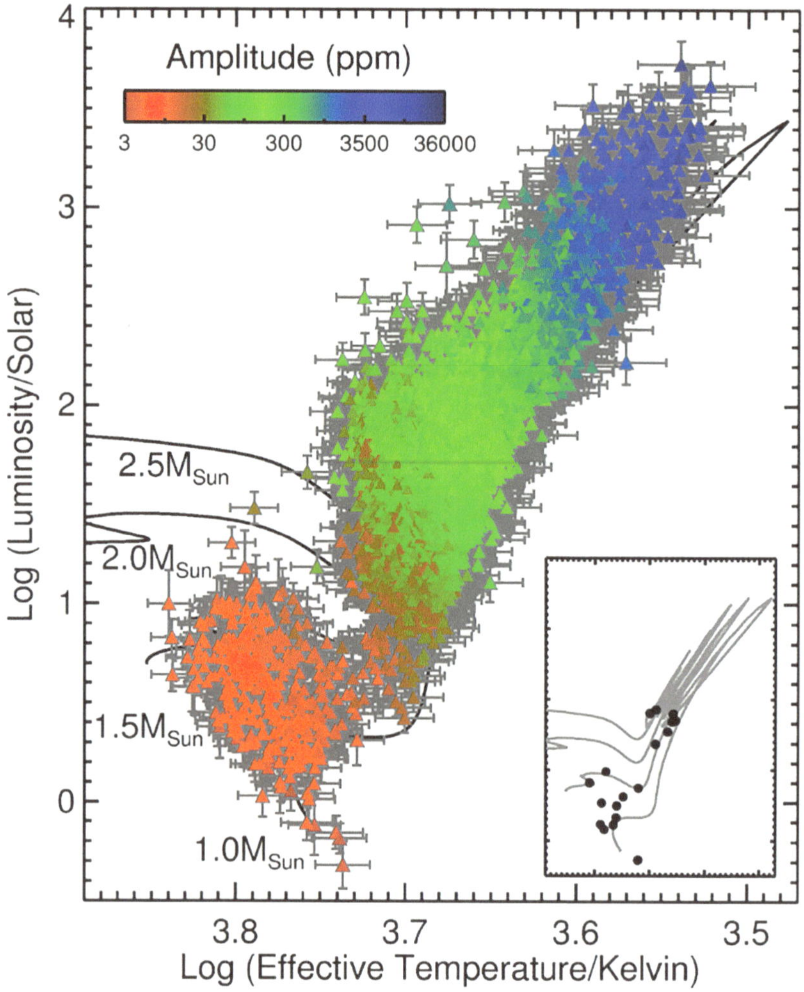

Figure 2.2. An H–R diagram showing the state of asteroseismic detections before (inset) and after the Kepler mission. Select evolutionary tracks are indicated, and the coloring indicates the mean amplitude of the observed oscillations. Reprinted with permission from Huber (2016).

transformed is small, of the order 10^{-5} L_*, even in cases where pulsation amplitudes are large, such as in Cepheids (Mundprecht et al. 2013). Consider the Cepheid light and radial velocity curves shown in Figure 2.3. This object, OGLE-LMC-CEP0227, was the first Cepheid detected in a well detached, double-lined eclipsing binary (Pietrzyński et al. 2010), giving an independent mass estimate for it of $3.98 \pm 0.29 M_\odot$, and a pulsational period of $P_{\rm pul} = 3.797086 \pm 0.000011$d.

Eddington (1941) suggested that oscillations in the classical instability strip are driven by a stellar implementation of a classical thermodynamic heat engine. He envisioned a situation where the opacity of a layer of gas inside the star increases as it is compressed. In that case, compression leads to increased blockage of energy flowing outward from the stellar interior, which in turn lifts the layer to larger radii.

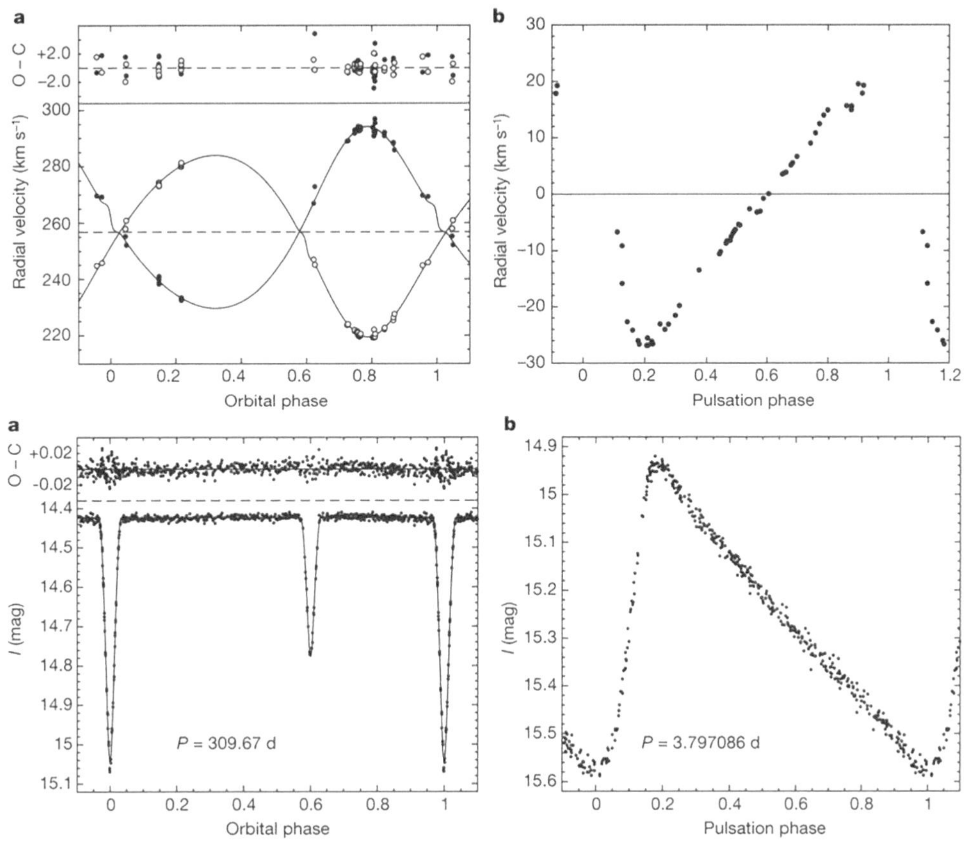

Figure 2.3. Top: The computed orbital radial velocity curves for both primary (filled) and secondary (open circles) components of the LMC-OGLE-CEP227 binary system. Residuals from the fit are shown above the fit in panel (a). Panel (b) shows the pulsational radial velocity curve; error bars are smaller than the points. Bottom: The orbital I-band light curve after removal of pulsational signals; the fit is overplotted and residuals are shown above. The phase-folded light curve is shown on the right. Reprinted with permission from Pietrzyński et al. (2010).

The resulting expansion leads to decreased opacity, the escape of the trapped energy, and the return of the later to its equilibrium position so the cycle can repeat itself. The basic mechanism is called the *kappa mechanism*, for its dependence on opacity κ (Baker & Kippenhahn 1962).

In order for the kappa mechanism to function, the layer must be one in which opacity increases when the gas is compressed, a circumstance which can occur in partial ionization zones. Here the ionization state is highly sensitive to temperature, so a small increase in temperature leads to a large change in the ionization state. Compression then does PdV work on the gas, but in these layers some of that work goes into ionizing the gas rather than raising the temperature, so the density rises while the temperature essentially doesn't. Since Kramer's-style opacity laws have $\kappa \sim \rho T^{-3.5}$, this implies that the opacity increases with compression, just as needed.

This dependence on ionization state leads to the more general term kappa-gamma mechanism for the process, where gamma (γ) is a thermodynamic quantity that varies with ionization state.

Where in the star does this happen? The most important partial ionization zones are those of the most common elements, hydrogen and helium. For hydrogen, with its single electron, there is of course only one zone, occurring at a temperature of about 10^4 K, while helium has two, the first (He I $\rightarrow$ He II) at 1.5×10^4 K and the second (He II $\rightarrow$ He III) at 4×10^4 K. This deeper second ionization zone is believed to be the driving mechanism for pulsations in classical Cepheids, such as shown in Figure 2.3. However, it's not enough for the zones to simply exist; if it were, all the stars in the HR diagram would display large-amplitude pulsations! Instead, the instability strip has boundaries at the low- and high-temperature ends. This is because the radial location of the ionization zones in the star is a function of the mass of the star. For high-mass stars, the ionization zones are relatively close to the stellar surface and contain little mass because the density there is too low, making it difficult for them to drive large oscillations. At low masses, the outer layer of the star is convective, providing an alternate means of heat transport and making mechanisms dependent on opacity increases ineffective.

For stars with masses lying above the instability strip, other mechanisms are required. Partial ionization zones due to iron and iron-like elements lie at temperatures near 2×10^5 K, while similar bumps due to carbon and oxygen occur at even higher temperatures. Iron and iron-like oscillation zones appear to be responsible for oscillations in OB stars such as β Cep and *SPB* stars, while C and O zones appear to power helium-rich subdwarfs and some white dwarf oscillations.

There is an alternative mechanism that can operate in the stellar core. Here, compression leads to an increase in both temperature and density, and therefore to the nuclear energy generation rate, which in turn causes expansion of the core. Expansion leads to a drop in density and temperature and return to equilibrium so the cycle can repeat. However, the amplitude of these oscillations appears to be too small to produce effects visible at the stellar photosphere, though the mechanism may operate in a meaningful way for the highest-mass stars.

It's worth noting that stellar oscillations can also be excited by means *external* to the star! For example, binary stars in eccentric orbits can be tidally excited if their resonance frequencies are close to the orbital frequency or its harmonics. An example of such a system, KOI-54, which has an orbital period of 41.805d, is shown in Figure 2.4. Here, the complex pattern visible in the light curve is due to beating between the 90th and 91st harmonics of the orbital frequency, though contributions from nearly 30 additional harmonics are also present.

2.2 Lower Main Sequence Stars

Lower main sequence stars are characterized by an internal structure consisting of an inner radiative core in which nuclear reactions occur, surrounded by a convective envelope. These stars typically have internal and surface magnetic fields, which

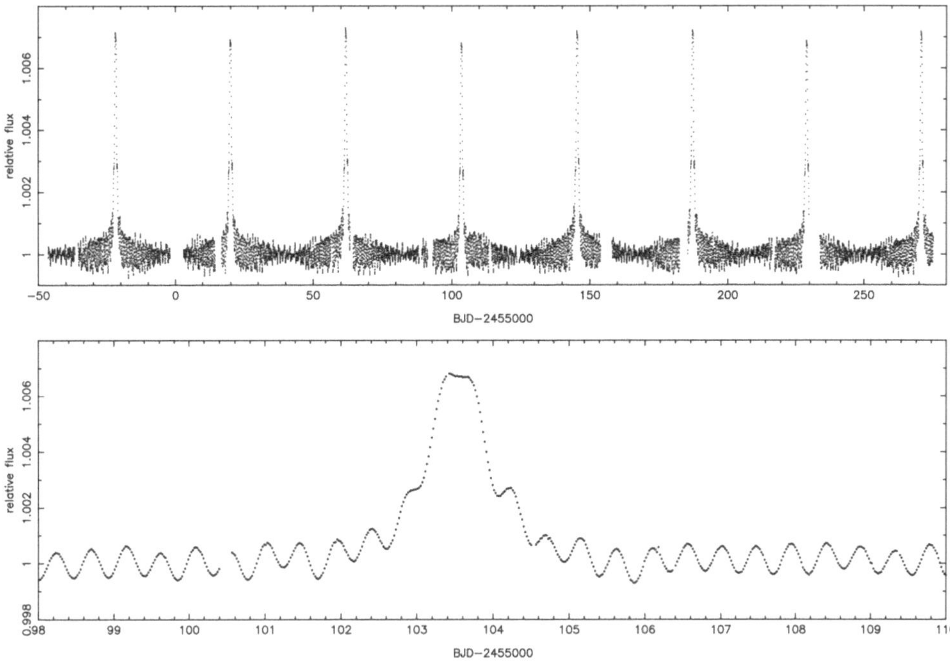

Figure 2.4. Top: The detrended and normalized Kepler light curve of KOI-54. The system has an orbital period of 41.805d, corresponding to a frequency of 0.023921d^{-1}, and the complex pattern of periodic brightenings visible is primarily due to excitation of the 90th (2.1529d^{-1}) and 91st (2.1768d^{-1}) harmonics of that orbital frequency. Bottom: A detailed view of a brightening event. Reprinted with permission from Welsh et al. (2011).

appear to be generated by a magnetic dynamo located in the overshoot region at the base of the outer convection zone.

Of course, the archetypical lower main sequence star is the Sun, and helioseismology has developed into a highly sophisticated and much-used tool since its beginnings in the 1960s (Basu 2016). In this case, our ability to resolve the solar photosphere enables us to detect and characterize modes with small wavelengths (high degree, l, see Section 3.7). Analysis of these oscillations has enabled researchers to test detailed solar models as well as to infer the 'average' radial structure of the Sun along with both polar and azimuthal departures from that structure. Asteroseismology allows the direct measurement of the internal sound speed profile, and from that inference of the internal density profile and rotation rate, as well as tests of models for convection, radiation opacity and chemical abundances within the Sun. Access to oscillations with short wavelengths also allows local helioseismic "imaging," which in turn enables mapping of flow fields around solar active regions (Figure 2.5).

The Sun shows a wide range of global oscillations with frequencies predominantly in the 2.5–4.5 mHz range, which early on became known as the so-called "five-

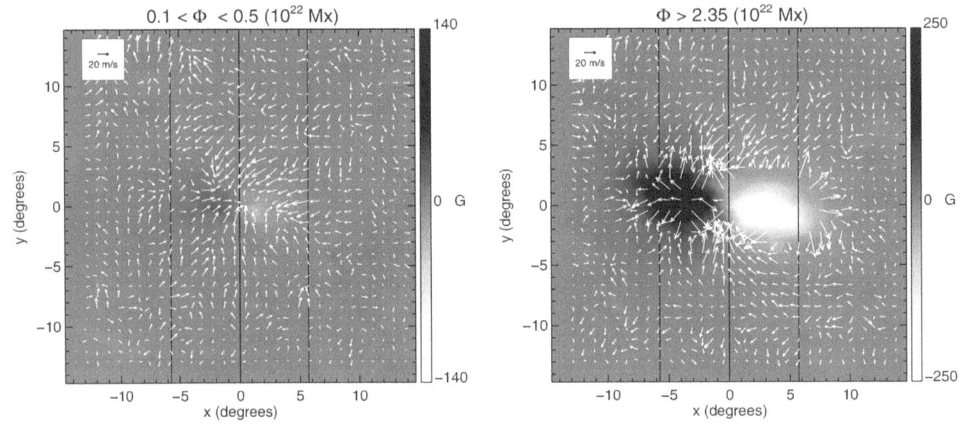

Figure 2.5. Near-surface horizontal flow fields (white arrows) derived from helioseismology, after averaging over active regions. The background grey scale shows the average line-of-sight field. Left: Average over active regions with Φ_B between 10^{21} and 5×10^{21}Mx. Right: Average over the largest active regions (>2.35×10^{22}Mx). Reprinted with permission from Birch (2020).

minute oscillations." The oscillations are visible in both velocity (*via* Doppler shifts in spectral lines) and in overall luminosity, though the amplitudes in both are quite small – the appropriate units are cm s^{-1} in velocity and parts-per-million (ppm) in luminosity. A large family of cool stars, both on the lower main sequence and the subgiant and giant branches, display oscillations which are similar in nature to those seen in the Sun, and we generically refer to these stars as "solar-like oscillators." Unlike more massive stars, oscillations in these stars are not driven by a heat engine and excited by the κ mechanism. Instead, they are due to the dissipation of turbulent energy arising from convection in the other layers of the star. This means that the excitation is stochastic rather than reflecting large-scale organization, and this lack of resonance implies that oscillation amplitudes are low, as we have seen above.

However, a characteristic of oscillations in solar-like oscillators that is generally absent in higher-mass stars is the observed pattern of frequencies present. Figure 2.6 shows the frequency spectrum for the solar-like star 16 Cyg A based on 4 years of Kepler photometry; presently this is the highest-quality seismic data available for any star other than the Sun. The spectrum shows excited frequencies confined to a fairly narrow range, with amplitudes roughly following a Gaussian envelope centered around a frequency of peak power, or ν_{max}. The frequencies themselves show a nearly regular separation between peaks, so that we can observationally define a large frequency separation, or $\Delta\nu$. In addition, both ν_{max} and $\Delta\nu$ are related to the stellar mass and radius, so provide a tool for the direct determination of these fundamental physical properties (Section 7.1).

Red giants in particular represent the evolved stages of moderate-mass stars like the Sun, once immediately-accessible hydrogen in the core becomes inaccessible to fusion reactions. The outer layers of the star expand and cool, leading to extremely deep outer convection zones. From an asteroseismic perspective, these stars are

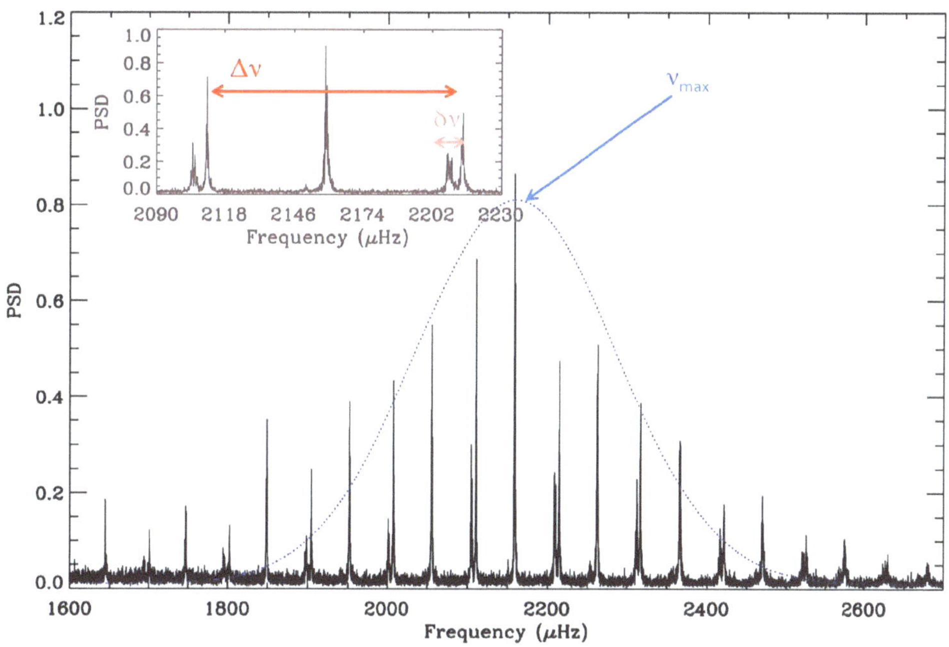

Figure 2.6. Power spectrum density (in arbitrary units) of the Kepler target 16 Cyg A. The blue dotted line represents the Gaussian fit to obtain the frequency at maximum power, while the inset above shows the large frequency separation, $\Delta\nu$, between two consecutive modes of angular degree and the small frequency separation $\delta\nu$. Reprinted with permission from García & Ballot (2019).

characterized by increased oscillation amplitudes and decreased frequencies as they ascend the red giant branch (see Figure 2.7 and discussion below).

2.3 Pre-MS Stars

Stars form from giant molecular clouds, passing through a phase of disk-mediated accretion and thermal contraction (on the Kelvin–Helmholtz timescale) before hydrogen fusion begins and they settle down on the Zero Age Main Sequence (ZAMS). While the general outline of the process has been understood for some time (Henyey et al. 1955; Hayashi 1961; Iben 1965), details remain somewhat obscure, and the simple picture in which stars smoothly move along the Hayashi and Henyey tracks, accreting steadily, until joining the main sequence, is clearly overly simplistic. Figure 2.8 illustrates the differences in stellar evolution tracks to the main sequence between older and newer models.

The internal structure of stars undergoing time-dependent disk-mediated accretion varies in ways which we currently do not understand well. Results from models make it clear that PMS stars with different internal structures, deriving from differing accretion histories, can nonetheless display very similar spectroscopic signatures, making traditional observational tools less useful for understanding the internal structure at times before the new star converges with the ZAMS.

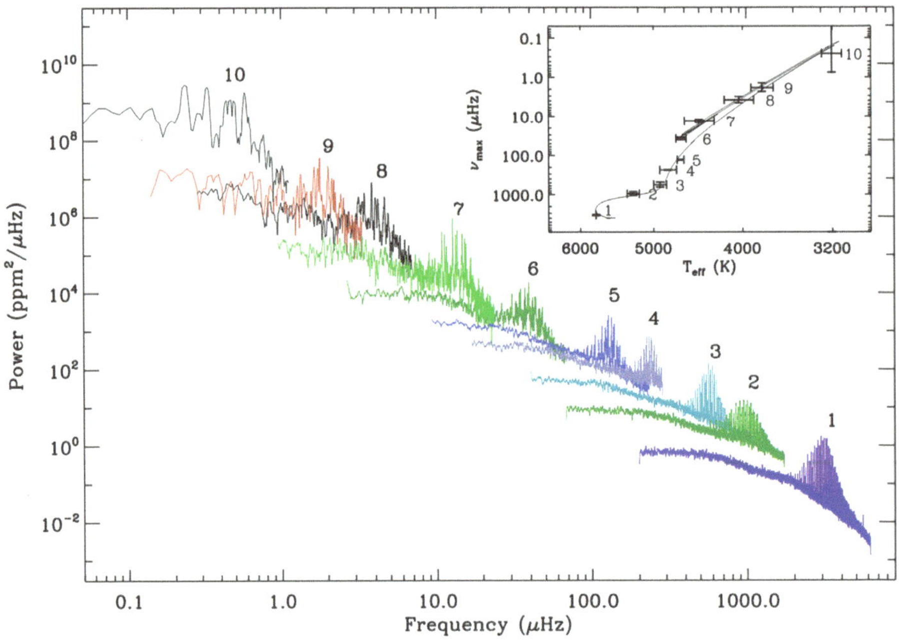

Figure 2.7. A sequence of stellar oscillation spectra starting from a main sequence star, the Sun (#1), and ranging up to the tip of the red giant branch. Star #6 is a red clump star. The inset represents a seismic HR diagram, where the solid line is the evolutionary track of a one solar-mass stellar model computed using MESA (Paxton et al. 2015), with the position of each star noted. The 1σ error bars in $\nu_{\max}$ are multiplied by a factor of 10 for visibility. Reprinted with permission from Jackiewicz (2021).

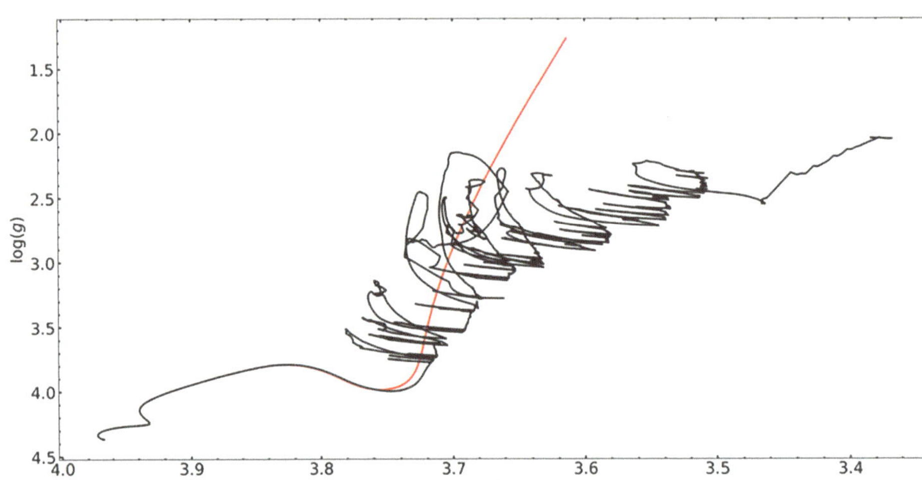

Figure 2.8. Kiel diagrams of a modern pre-main sequence model with time-dependent disk-mediated accretion (black) and a "classical" model following the Hayashi–Henyey tracks. Broadly speaking, the tracks are similar, but the evolution of the disk-mediated model follows a far more complex route to the ZAMS. Reprinted with permission from Zwintz & Steindl (2022).

Asteroseismology promises to be a valuable tool for probing the internal structure of these protostars, though the active nature of many of them along with the presence of shrouding material complicates oscillation detection. To date, most detections have come from stars more massive than the Sun, and over 100 PMS stars have been found to show δ Scu, SPB, γ Dor, and hybrid-type oscillations. Oscillations excited by tidal perturbations in binary systems, along with solar-like oscillations, have only single candidate objects so far (Zwintz & Steindl 2022). Figure 2.9 shows oscillations from the intermediate-mass Herbig Ae star HD 142666.

2.4 Evolved Stars

Stars with masses below roughly $1.5M_{\odot}$ undergo an increase in helium content, and hence in mean molecular weight μ, over time. This increase leads to contraction of the core and a corresponding increase in its temperature and in the energy generation rate ε. Since the core is radiative, the core exhausts hydrogen from the inside out, producing an inert center of helium "ash" surrounded by a hydrogen-burning shell. More massive stars have convective cores, and the mixing produces a uniform profile of hydrogen exhaustion within the core. As with lower-mass stars, this leads to shell-burning, but the onset is more sudden.

The onset of shell burning occurs simultaneously with core contraction and envelope expansion (the so-called "mirror principle"; Hekker 2020). The deep envelope is convective, and overlies a small dense radiative core, partly supported by electron degeneracy pressure. The deepening convective envelope dredges up material from earlier evolutionary phases, modifying the mean molecular weight profile of the star as it moves onto the red giant branch.

The core continues to grow in both mass and temperature until the latter reaches $\sim 10^8$ K, at which point helium fusion in the core begins via the triple-alpha process. In lower-mass stars, the core is supported primarily by electron degeneracy pressure, which is temperature-independent, so the onset of helium burning occurs in a context where the normal thermostatic controls are absent. This means that helium ignition is a dramatic event, the "helium flash."[2] In more massive stars ($M \geqslant 2M_{\odot}$), degeneracy is a less important contributor to core pressure, so helium ignition is a more sedate event.

After helium ignition, the star is producing energy via two separate fusion processes: He $\rightarrow C/O$ in the core and H $\rightarrow$ He in a shell surrounding the core. The shell expands, while the envelope mirrors that expansion with contraction; observationally, the star is in the red clump.[3] Convective overshoot likely drives significant mixing among the different compositional layers.

Eventually helium is exhausted in the core, which becomes an inert C/O "ash" region continuing to contract, while He fusion continues in a second shell

---

[2] Models indicate that the flash is likely multiple events rather than a single one.

[3] The "clump" nature arises because at this point the variation of luminosity as a function of stellar mass is small, and because the star spends a significant period of time in this phase of its evolution compared to other points along the RGB.

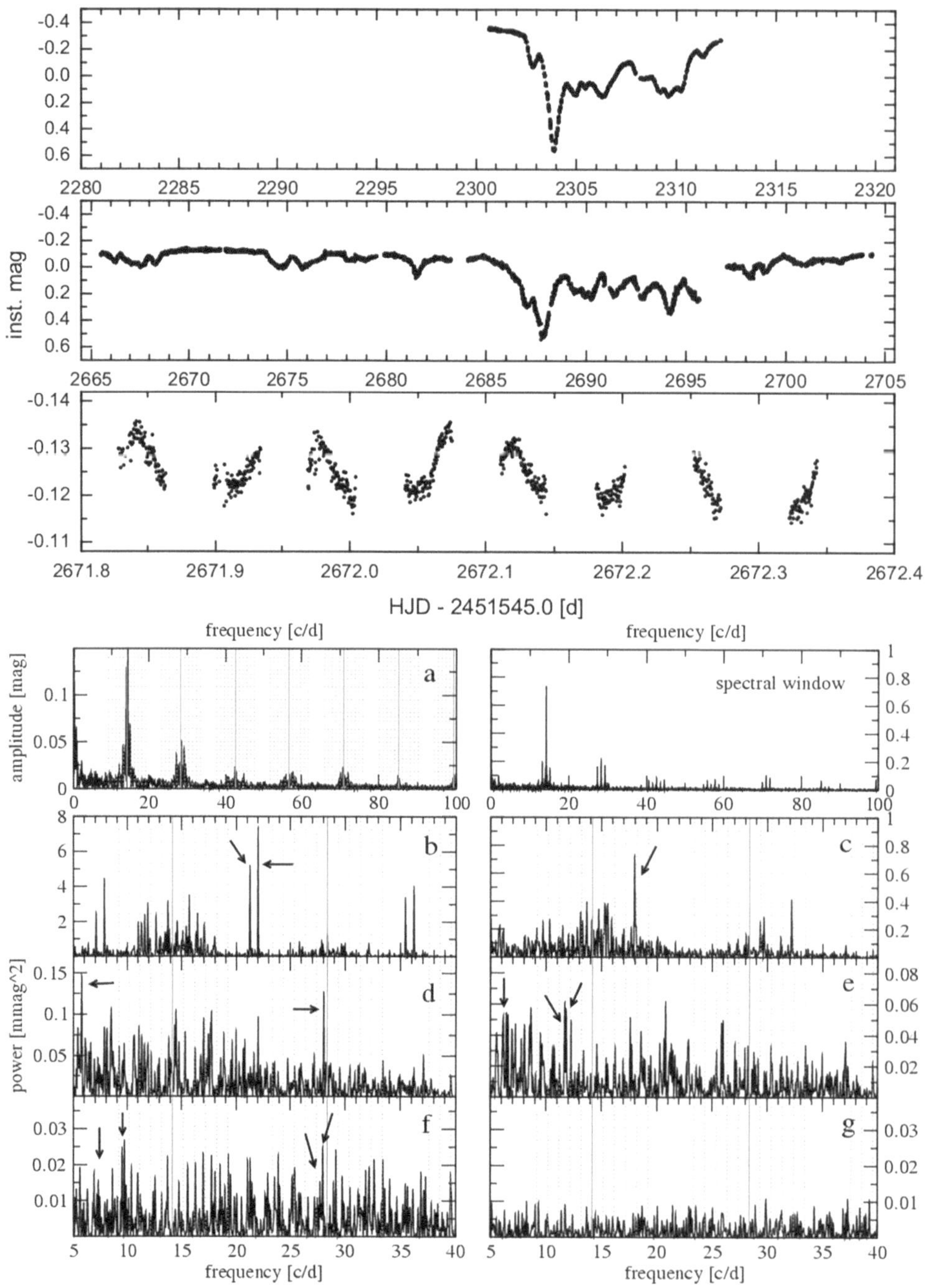

Figure 2.9. Top: MOST time series photometry of the pre-main sequence star HD 142666 from 2006 and 2007. The lower panel zooms into a small section to show details. Bottom: Frequency analyses of the 2006 photometry. Identified pulsation frequencies are indicated with arrows, while the solid (and dotted) gray lines show the satellite orbital frequency and harmonics (and sidelobes). Reprinted with permission from Zwintz et al. (2009).

immediately surrounding the core. Observationally, the star now ascends the asymptotic giant branch (AGB), an ascent which is punctuated by the onset of multiple thermal pulses, driving additional mixing which can move fusion products such as C all the way to the surface, where they become visible to spectroscopy. Pulses also expel envelope material into space, helping to produce a planetary nebula. Higher-mass stars can reach temperatures sufficient to fuse C and O in their cores, leading to a more complex "onion" structure, but the end result is similar: a cooling, dense white dwarf supported by degeneracy pressure.

Red giants have convective outer layers, which leads us to anticipate p-mode oscillations in the envelope. In fact, this is what is observed, but the typical timescale is longer, as the free-fall time is related to the mean stellar density, and red giants have significantly lower densities than main sequence stars. Figure 2.10 shows the effects: stellar densities are lower toward the top of the figure, leading to lower frequency oscillations.

In the Sun, the radiative core is expected to undergo g-mode oscillations, which have low frequencies and which are sufficiently damped in the envelope so as to have as-yet-undetectable signals at the surface.[4] Red giant stars, with their radiative cores, also have g-mode oscillations. However, the low frequencies of p-mode oscillations has another effect as well, because the periods of core and envelope oscillations become similar to one another. This leads to interactions which give the oscillations a mixed character, combining g-mode and p-mode behavior, where the dominant behavior is determined by the coupling strength between the two (or more!) oscillation cavities. In practice, we can't observe the g-mode oscillation periods directly, but we can *infer* them from their effect on the p-modes, which we *can* observe at the surface layers.

Red giants have complex evolutionary tracks and structure, and traditional astronomical observational tools can be hampered for their analysis by the fact that a range of differing internal structures produce overall surface diagnostics (as one might see in the HR or Kiel diagram) which are in fact quite similar. Figure 2.11 illustrates one way in which asteroseismology can break this degeneracy, by allowing us to distinguish unambiguously between H-shell burning stars, with period spacings of approximately 50 s, and stars which are fusing He in their cores, which have period spacings of 100–300 s, despite significant overlaps in the measured large separation $\Delta\nu$. The access these mixed modes grant us to seismology of the deep interior and core makes red giants in many ways an ideal application for asteroseismology.

2.5 White Dwarfs

White dwarfs, which typically have masses between 0.15 and $1.25M_\odot$, are the end state for most stars with initial masses below $9\text{–}11M_\odot$. With radii $\sim 0.01R_\odot$, they are extremely dense compact objects supported by electron degeneracy pressure. As stellar end products, their composition is dominated by C/O, though low-mass white dwarfs contain substantial He and the most massive white dwarfs likely contain significant quantities of Ne and Mg, and composition is rendered somewhat

[4] Models suggest surface radial velocity signals on the order of mm s^{-1}.

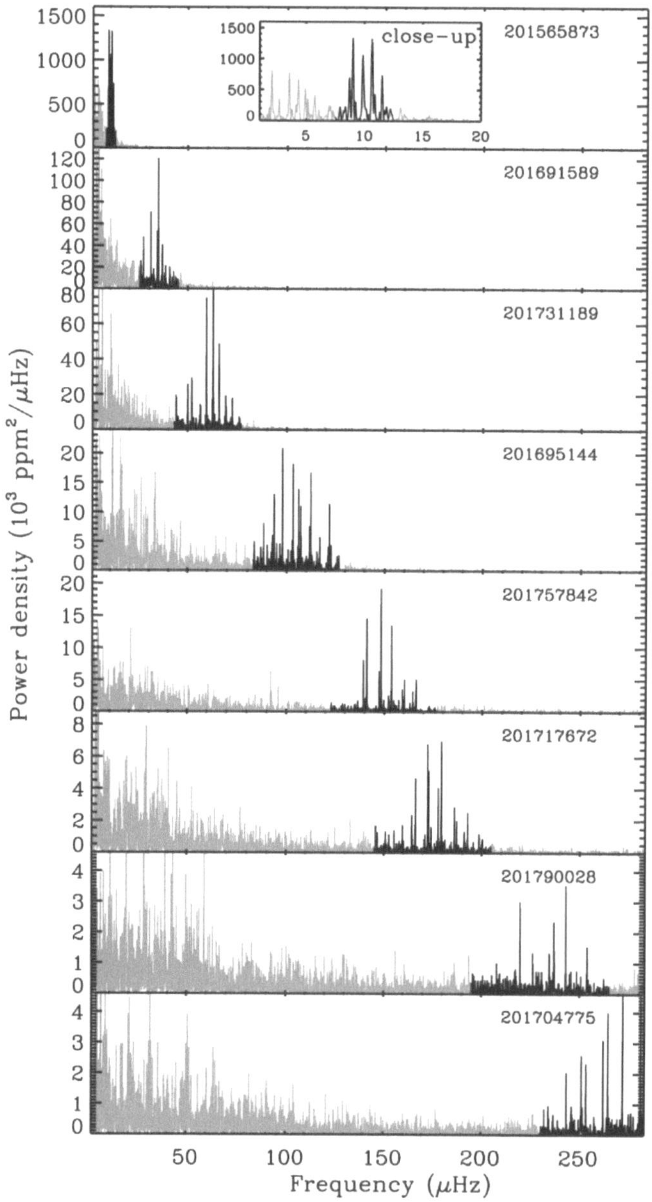

Figure 2.10. Power spectra of selected red giants (labeled by EPIC ID) representing the range in $\nu_{\rm max}$ detectable from K2 long-cadence data. The stars are ordered from most luminous (lowest oscillation frequencies of about 10 μHz) at the top, to the least luminous (highest oscillation frequency of about 270 μ Hz) at the bottom. The frequency range dominated by the oscillations is shown in black. Reprinted with permission from Stello et al. (2015).

nonuniform through gravitational settling, so most have thin H/He atmospheres surrounding their degenerate cores. Since they no longer produce significant energy, the effective temperature of white dwarfs cools predictably over time, from

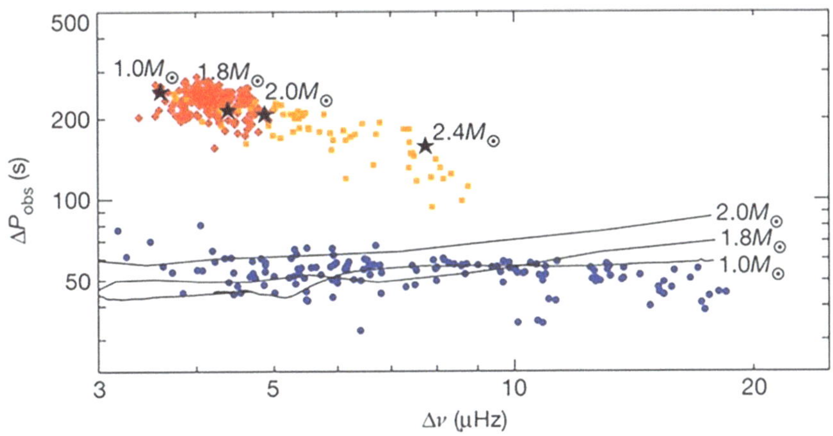

Figure 2.11. Period spacings plotted against the p-mode large frequency separation for a selection of red giants observed with Kepler. The stars divide into two clear groups, with blue circles indicating hydrogen-shell-burning giants, and the remainder helium-core-burning stars (red diamonds and orange squares). Solid lines show average observable period spacings from models of hydrogen-shell-burning giants on the red giant branch as they evolve from right to left, calculated from the central three modes in the $l = 1$ clusters. The black stars show theoretical period spacings calculated in the same way, for four models of helium-core-burning stars that are midway through that phase (core helium fraction 50%). Reprinted with permission from Bedding et al. (2011).

formation ($T_e \sim 2 \times 10^5$ K) to as little as $T_e = 4000$ K; combined with their small radii this implies a wide range of luminosities, with most being quite faint compared to main sequence stars. The age and long lifetimes of these objects make them useful as probes of stellar evolution and Galactic history.

From an asteroseismic perspective, white dwarfs are valuable because they display non-radial g-mode oscillations with typical periods of 10^2–10^3 s. At optical and near-optical wavelengths, the oscillation amplitudes can be quite large (up to ~0.4 mag), and probe the structure of the entire star. The first oscillating white dwarf was discovered serendipitously by Landolt (1968), and the large oscillation amplitudes made study using ground-based telescopes possible using coordinated observations from a network of telescopes (the "Whole Earth Telescope," or WET) for decades prior to the availability of space-based platforms. Figure 2.12 shows an example of such ground-based observations. From early on, it became clear that light curve complexity ranged from simple variations characterized by only a single period to complex variability containing multiple oscillation frequencies modified by nonlinear features and interactions with other physical phenomena in the outer layer and atmosphere of the star (Córsico 2020).

Figure 2.13 shows the different types of white dwarfs on a Kiel diagram with some representative evolutionary tracks. The main types include DAV (or ZZ Ceti type) stars, which have H-dominated atmospheres, DBV (or V777 Her type) stars, with He-dominated atmospheres, and GW Vir stars, with more complex atmospheres showing C and O as well as He. The figure also shows the location of the extremely low-mass variable (ELMV) white dwarfs, an oscillating subset with masses under about $0.45M_\odot$, believed to have He cores.

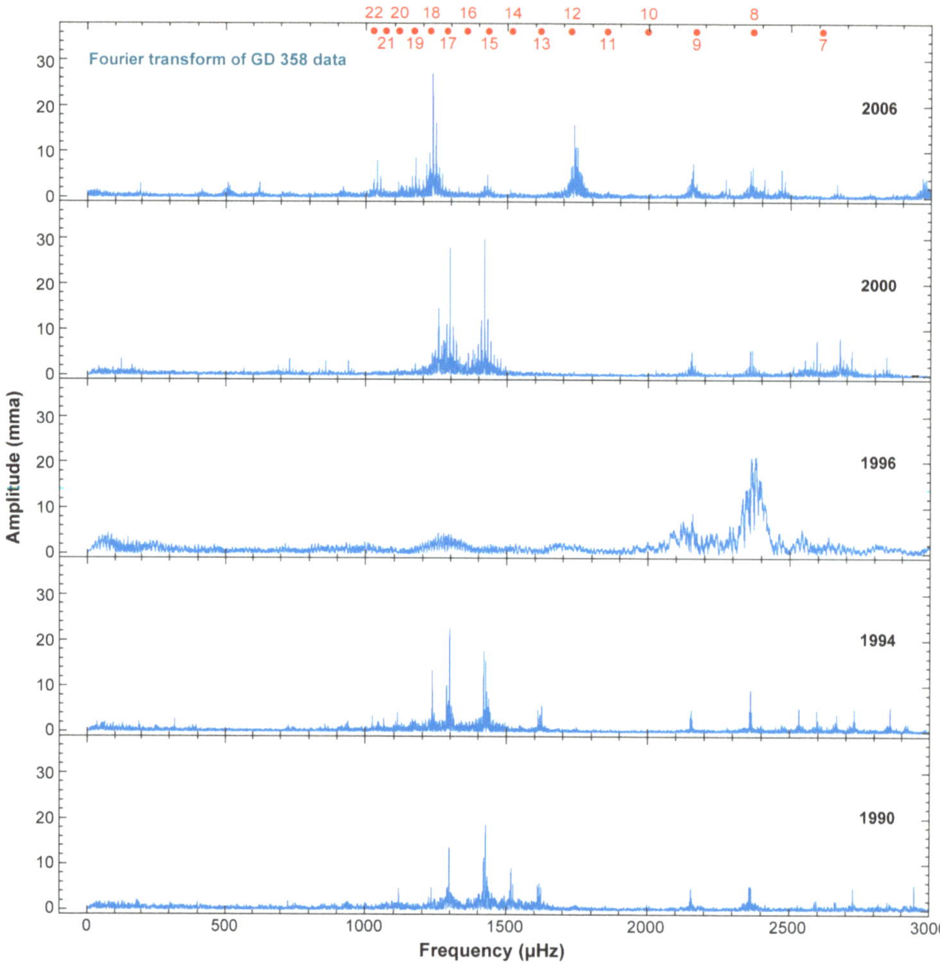

Figure 2.12. Amplitude spectra of the combination of several Whole Earth Telescope (WET) campaigns on the DBV white dwarf GD 358. The red marks on the top indicate the expected location of modes following an asymptotic ΔP analysis. Some small alias peaks remain, due to the existence of gaps in the light curves. Reprinted with permission from Winget & Kepler (2008).

The driving mechanism for oscillations in white dwarfs depends on the partial ionization of their outer layers. The underlying radiative layers heat the base of the ionization zone, and resulting opacity increases act as a driving mechanism in a process known as "convective driving" (Brickhill 1991). For DAV white dwarfs, hydrogen ionization provides the basis for the driving, while for the hotter DBV white dwarfs it is helium ionization, and finally in the hottest GW Vir stars it is K-shell electrons of carbon and oxygen. White dwarf oscillations are non-radial g-mode pulsations, with buoyancy acting as the restoring force. Unlike solar-type oscillators, not all resonant frequencies in white dwarfs are excited to large amplitudes, though frequently evenly-spaced period structure is visible, which can be translated into information about their internal structure and rotation (Bell 2025).

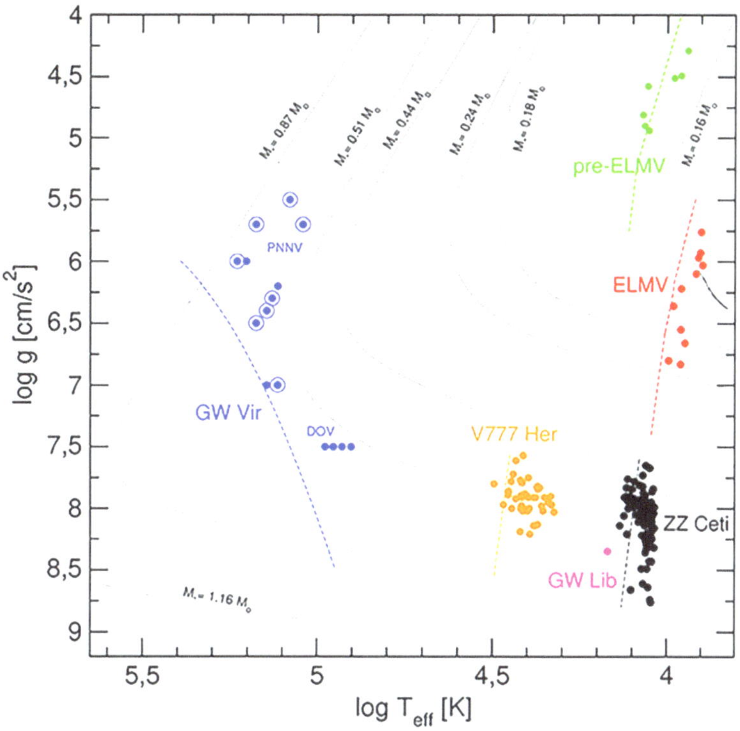

Figure 2.13. Different families of confirmed pulsating white dwarf and pre-white dwarf stars (circles of different colors) in the $\log T_e - \log g$ plane. GW Vir stars indicated with blue circles surrounded by blue circumferences are PNNVs. The location of the prototypical GW Lib object is indicated by a magenta dot. Two post-VLTP (Very Late Thermal Pulse) evolutionary tracks for H-deficient white dwarfs (0.51 and $0.87 M_\odot$; Miller Bertolami & Althaus 2006), four evolutionary tracks of low-mass He-core H-rich white dwarfs (0.16, 0.18, 0.24, and $0.44 M_\odot$; Althaus et al. 2013), and one evolutionary track for ultra-massive H-rich white dwarfs ($1.16 M_\odot$; Camisassa et al. 2019) are included for reference. Dashed lines correspond to the location of the theoretical hot boundary of the different white dwarf instability strips. Reprinted with permission from Córsico (2020).

2.6 Classical Pulsators

So-called classical pulsators have large amplitudes because they lie within the instability strip on the HR diagram. The most famous examples are the δ Cepheid stars (Type I and Type II Cepheids) and the RR Lyrae stars, though there are additional subtypes as well (see Figure 4.8). All pulsate[5] in low-order radial modes, generally either the fundamental or the first overtone, and are driven by the κ mechanism. In the case of the Cepheids, the mechanism is based on the second ionization of helium. Type I "Classical" Cepheids are younger and more massive

[5] The correct usage of *pulsate* versus *oscillate* to describe periodic intrinsic variability is somewhat obscure. I will use the former to describe variability due to low-order radial modes, particularly when the amplitude of the variability is large.

(4–20$M_\odot$ versus ~0.5–0.8$M_\odot$) than Type II Cepheids, and are unsurprisingly members of population I while Type II Cepheids are in population II. Typical periods for both classes are from 1–100 d, and both show period–luminosity relationships, though Type I Cepheids are approximately 1.5 mag brighter than Type II Cepheids with the same period. Type II Cepheids are broken into subclasses, including W Virginis, RV Tauri, and BL Her stars based primarily on pulsation period. RR Lyrae stars are also lower-mass (~0.5$M_\odot$) older (Population II) stars, and have a similar driving mechanism to Cepheids, but lie on the horizontal branch. Both are large-amplitude, single-mode radial oscillators with periods easily accessible from both the ground and space, and possess period–luminosity relationships, making them excellent distance indicators.

Even without a period–luminosity relationship, the large amplitude of the pulsations in these stars makes use of the Baade–Wesselink method usable as an independent distance determination method. In this approach, if we treat a pulsating star as a blackbody, then we can write the flux as

$$F = \frac{L}{4\pi D^2} = \frac{4\pi R^2 \sigma T_e^4}{4\pi D^2} = \frac{R^2 \sigma T_e^4}{D^2}, \tag{2.1}$$

where D is distance to the star. Taking logarithms gives

$$\log F = 2\log R + 4\log T_e + \log \sigma - 2\log D \tag{2.2}$$

and differentiating with respect to time

$$\frac{d}{dt}\log F = 2\frac{d}{dt}\log R + 4\frac{d}{dt}\log T_e = \frac{2}{R}\frac{dR}{dt} + 4\frac{d}{dt}\log T_e \tag{2.3}$$

We leave flux F in logarithmic terms here in recognition of the logarithmic nature of the astronomical magnitude scale. With the help of time-resolved high-resolution spectroscopy to determine the photospheric expansion/contraction velocity $v = dR/dt$, we can solve for R, which in turn allows solution for D from the original expression.

While Cepheids and RR Lyrae stars lie where the instability strip intersects the evolutionary paths of more evolved stars, *rapidly oscillating Ap* stars and δ Scuti stars lie on or near where the strip crosses the main sequence. roAp stars oscillate in high-order, low-degree non-radial p-modes, with periods of 5–30 min and optical amplitudes of mmag. However, pulsation amplitude is modulated by the stellar rotation period, leading to an *oblique pulsator model* for these stars, where the pulsation axis is aligned with the magnetic rather than the rotational axis. The driving mechanism for roAp stars is not yet known. δ Scuti stars are somewhat less massive, with spectral types A–F, periods of hours, and optical pulsational amplitudes as large as ~0.5 mag or more.

References

Althaus, L. G., Miller Bertolami, M. M., & Córsico, A. H. 2013, A&A, 557, A19

Baker, N., & Kippenhahn, R. 1962, Zeitschrift fuer Astrophysik, 54, 114

Basu, S. 2016, LRSP, 13, 2
Bedding, T. R., Butler, R. P., Kjeldsen, H., et al. 2001, ApJL, 549, L105
Bedding, T. R., Mosser, B., Huber, D., et al. 2011, Natur, 471, 608
Bell, K. J. 2025, arXiv:2502.13258
Birch, A. C. 2020, Astrophysics and Space Science Proc., Vol. 57, Dynamics of the Sun and Stars; Honoring the Life and Work of Michael ThompsonJ. ed. M. J. P. F. G. Monteiro, R. A. García, J. Christensen-Dalsgaard, & S. W. McIntosh (Berlin: Springer International Publishing) 91
Brickhill, A. J. 1991, MNRAS, 251, 673
Brown, T. M., Gilliland, R. L., Noyes, R. W., & Ramsey, L. W. 1991, ApJ, 368, 599
Bruntt, H., & Buzasi, D. L. 2006, MmSAI, 77, 278
Bruntt, H., Kjeldsen, H., Buzasi, D. L., & Bedding, T. R. 2005, ApJ, 633, 440
Buey, J. T., Auvergne, M., Vuillemin, A., & Epstein, G. 1997, PASP, 109, 140
Buzasi, D. 2002, Conf. Ser. 259, IAU Colloq. 185: Radial and Nonradial Pulsationsn as Probes of Stellar Physics ed. C. Aerts, T. R. Bedding, & J. Christensen-Dalsgaard (San Francisco, CA: ASP) 616
Camisassa, M. E., Althaus, L. G., Córsico, A. H., et al. 2019, A&A, 625, A87
Chaplin, W. J., Kjeldsen, H., Christensen-Dalsgaard, J., et al. 2011, Science, 332, 213
Córsico, A. H. 2020, FrASS, 7, 47
Eddington, A. S. 1941, MNRAS, 101, 182
Edmonds, P. D., & Gilliland, R. L. 1996, ApJL, 464, L157
García, R. A., & Ballot, J. 2019, LRSP, 16, 4
Hatt, E., Nielsen, M. B., Chaplin, W. J., et al. 2023, A&A, 669, A67
Hayashi, C. 1961, PASJ, 13, 450
Hekker, S. 2020, FrASS, 7, 3
Hekker, S., Kallinger, T., Baudin, F., et al. 2009, A&A, 506, 465
Henyey, L. G., Lelevier, R., & Levée, R. D. 1955, PASP, 67, 154
Hoffleit, D. 1997, JAVSO, 25, 115
Huber, D. 2016, arXiv:1604.07442
Iben, I. 1965, ApJ, 141, 993
Jackiewicz, J. 2021, FrASS, 7, 102
Kjeldsen, H., Bedding, T. R., Frandsen, S., & dall, T. H. 1999, MNRAS, 303, 579
Landolt, A. U. 1968, ApJ, 153, 151
Martić, M., Schmitt, J., Lebrun, J. C., et al. 1999, A&A, 351, 993
Matthews, J. M. 1998, Structure and Dynamics of the Interior of the Sun and Sun-like Stars SOHO 6/GONG 98 Workshop Abstract Vol. 418, ed. S. Korzennik; Paris: European Space Agency) 395
Matthews, J. M. 2007, CoAst, 150, 333
Michel, E., Baglin, A., Auvergne, M., et al. 2008, Sci, 322, 558
Miller Bertolami, M. M., & Althaus, L. G. 2006, A&A, 454, 845
Moya, A., Barceló Forteza, S., Bonfanti, A., et al. 2018, A&A, 620, A203
Mundprecht, E., Muthsam, H. J., & Kupka, F. 2013, MNRAS, 435, 3191
Pápics, P. I. 2013, PhD Thesis, Instituut voor Sterrenkunde, KU Leuven
Paxton, B., Marchant, P., Schwab, J., et al. 2015, ApJS, 220, 15
Pietrzyński, G., Thompson, I. B., Gieren, W., et al. 2010, Natur, 468, 542

Schou, J., & Buzasi, D. L. 2001, SOHO 10/GONG 2000 Workshop: Helio- and Asteroseismology at the Dawn of the Millennium (Vol. 464, ed. A. Wilson, & P. L. Pallé; Paris: European Space Agency) 391
Stello, D., Bruntt, H., Preston, H., & Buzasi, D. 2008, ApJL, 674, L53
Stello, D., Huber, D., Sharma, S., et al. 2015, ApJL, 809, L3
Tarrant, N. J., Chaplin, W. J., Elsworth, Y., Spreckley, S. A., & Stevens, I. R. 2007, MNRAS, 382, L48
Walker, G., Matthews, J., Kuschnig, R., et al. 2003, PASP, 115, 1023
Weiss, W. W., Zwintz, K., Kuschnig, R., et al. 2021, Univ, 7, 199
Welsh, W. F., Orosz, J. A., Aerts, C., et al. 2011, ApJS, 197, 4
Winget, D. E., & Kepler, S. O. 2008, ARA&A, 46, 157
Zwintz, K., & Steindl, T. 2022, FrASS, 9, 914738
Zwintz, K., Kallinger, T., Guenther, D. B., et al. 2009, A&A, 494, 1031

Asteroseismology for the Nonspecialist

Derek L Buzasi

Chapter 3

The Physics of Stellar Oscillations

This chapter constitutes a fairly minimalist approach to the fundamental equations describing stellar oscillations. Considerably more detailed expositions can be found in, e.g., Gough (1993), Aerts et al. (2010), Chaplin & Miglio (2013), Pallé & Esteban (2014), Basu & Chaplin (2018), Aerts (2021), and Kurtz (2022) for stars in general, Bowman (2020) for high-mass stars, Garcia & Stello (2018) for solar-type stars, and Garcia & Stello (2018) for red giants.

3.1 Mathematical Preludes

Since stars are spherical objects (at least to first order), the natural coordinate system to use to describe them is *spherical polar coordinates*, in which the position of a point is characterized by (r, θ, ϕ). Here r is the distance from the origin (so has units of length, such as meters), θ is the *colatitude* or the *polar angle*, the angle measured from the polar axis, such that the equator of a sphere is at $\theta = 90° = \pi/2$ rad,[1] and ϕ is the *longitude*, measured in angular units from some (usually arbitrarily or conveniently chosen) zero point, such as the x-axis. This allows us to characterize the position of an arbitrary point by

$$\vec{A} = A_r \vec{e_r} + A_\theta \vec{e_\theta} + A_\phi \vec{e_\phi} \tag{3.1}$$

It is straightforward but tedious to show that the gradient operator in spherical polar coordinates can be written as:

$$\nabla \vec{A} = \frac{\partial A}{\partial r}\vec{e_r} + \frac{1}{r}\frac{\partial A}{\partial \theta}\vec{e_\theta} + \frac{1}{r \sin\theta}\frac{\partial A}{\partial \phi}\vec{e_\phi} \tag{3.2}$$

[1] There are alternate versions of the system, such that θ is the latitude instead, or where θ and ϕ trade names; neither is right or wrong, but it pays to be sure which version you're using!

doi:10.1088/2514-3433/ae03a0ch3

and the divergence operator as

$$\nabla \cdot A = \frac{1}{r^2}\frac{\partial}{\partial r}(r^2 A_r) + \frac{1}{r\sin\theta}\frac{\partial}{\partial\theta}(\sin\theta A_\theta) + \frac{1}{r\sin\theta}\frac{\partial A}{\partial\phi} \tag{3.3}$$

Together we can use these to construct the Laplacian operator

$$\nabla \cdot \nabla\, A \equiv \nabla^2\, A = \frac{1}{r^2}\frac{\partial}{\partial r}\left(r^2\frac{\partial v}{\partial r}\right) + \frac{1}{r^2\sin\theta}\frac{\partial A}{\partial\theta}(\sin\theta A_\theta) + \frac{1}{r^2\sin^2\theta}\frac{\partial^2 A}{\partial\phi^2} \tag{3.4}$$

Since, as we have seen, stars are (to first order at least!) self-gravitating spherical bodies, we can easily describe them from the perspective of Poisson's equation (see Equation (3.83) below). We will incorporate this into our description of oscillations below, but it's instructive first to write Laplace's equation

$$\nabla^2 A = 0 \tag{3.5}$$

and solve it in spherical polar coordinates, to see the form those solutions take. We won't go through this exercise in its full glorious detail (see any advanced text in electricity and magnetism or on introductory quantum mechanics), but note that the process makes use of separation of variables, where we assume that

$$A = R(r)\Theta(\theta)\Phi(\phi) \tag{3.6}$$

Inserting this into Equation (3.4) gives (after a little algebra)

$$\frac{1}{R}\frac{d}{dr}\left(r^2\frac{dR}{dr}\right) + \frac{1}{\Theta\sin\theta}\frac{d}{d\theta}\left(\sin\theta\frac{d\Theta}{d\theta}\right) + \frac{1}{\Phi\sin^2\theta}\frac{d^2\Phi}{d\phi^2} = 0 \tag{3.7}$$

Multiplying through by $\sin^2\theta$ allows us to separate the Φ portion, so

$$\frac{1}{\Phi}\frac{d^2\Phi}{d\phi^2} = -\frac{\sin^2\theta}{R}\frac{d}{dr}\left(r^2\frac{dR}{dr}\right) - \frac{\sin\theta}{\Theta}\frac{d}{d\theta}\left(\sin\theta\frac{d\Theta}{d\theta}\right) = -m^2 \tag{3.8}$$

Since both sides of the equality depend on different variables, and yet are equal, they must equal a constant. Here we call this *separation constant* $-m^2$, where m is an integer, which is chosen to be negative so that solutions for $\Phi(\phi)$ are periodic over $\phi \to \phi + 2\pi$, and the solution has the form $e^{\pm im\phi}$.

Now we can use this to rewrite Equation (3.7) as

$$\frac{1}{R}\frac{d}{dr}\left(r^2\frac{dR}{dr}\right) = -\frac{1}{\Theta\sin\theta}\frac{d}{d\theta}\left(\sin\theta\frac{d\Theta}{d\theta}\right) - \frac{m^2}{\sin^2\theta} = l(l+1) \tag{3.9}$$

Once again, both sides must equal a constant, which with the benefit of foreknowledge we write here as $l(l+1)$, where l is an integer. The radial equation is an Euler equation

$$r^2\frac{d^2R}{dr^2} + 2r\frac{dR}{dr} - l(l+1)R = 0 \tag{3.10}$$

which has solutions which are linear combinations of r^l and r^{-l-1}, while the solution to the Θ portion,

$$\frac{1}{\sin\theta}\frac{d}{d\theta}\left(\sin\theta\frac{d\Theta}{d\theta}\right)+\left[l(l+1)-\frac{m^2}{\sin^2\theta}\right]\Theta=0 \tag{3.11}$$

is the set of differential equations defining the associated *Legendre polynomials* $P_l^m(\cos\theta)$.

At this point, we introduce the *spherical harmonics*, which have the form

$$Y_l^m(\theta,\phi)=(-1)^m c_{lm}P_l^m(\cos\theta)e^{im\phi} \tag{3.12}$$

where the c_{lm} are just normalization factors, and the Y_l^m themselves are known as the *spherical harmonics*. Both l and m are integers, where $l = 0, 1, 2, \ldots$ and $m = -l, -l+1, \ldots, l-1, l$. Perhaps the greatest value of the spherical harmonics for our purposes is that they constitute a set of orthonormal basis functions on the sphere, which means that we can use them to construct **any** well-behaved function of θ and ϕ using linear combinations of spherical harmonics, so

$$f(\theta,\phi)=\sum_{l=0}^{\infty}\sum_{m=-l}^{l}a_{lm}Y_l^m(\theta,\phi) \tag{3.13}$$

where the a_{lm} are weighting factors. The spherical harmonics for specific values of l and m are simple to look up, but for illustrative purposes a sample of the first few look like this:

$$\begin{aligned}Y_0^0(\theta,\phi)&=\frac{1}{\sqrt{4\pi}}\\ Y_1^0&=\sqrt{\frac{3}{4\pi}}\cos\theta\\ Y_1^1&=-\sqrt{\frac{3}{8\pi}}\sin\theta e^{i\phi}\end{aligned} \tag{3.14}$$

There's a handy recursion relation available if you'd rather calculate further Y_l^m. In particular, we can use them to construct a general solution to Laplace's equation:

$$A(r,\theta,\phi)=\sum_{l=0}^{\infty}\sum_{m=-l}^{l}(a_{lm}r^l+b_{lm}r^{-l-1})Y_l^m(\theta,\phi) \tag{3.15}$$

We can rewrite the Laplacian using spherical harmonics as

$$\nabla^2[(a_{lm}r^l+b_{lm}r^{-l-1})Y_l^m(\theta,\phi)]=0 \tag{3.16}$$

which reduces to

$$-r^2\,\nabla^2\,Y_l^m(\theta,\phi)=l(l+1)Y_l^m(\theta,\phi), \tag{3.17}$$

The separability of the equation into r and (θ, ϕ) suggests that it might also be useful to break some operators up as well. Recall that our stellar models are one-dimensional, with physical quantities such as temperature and pressure functions of r only, and not depending on (θ, ϕ) at all. We can accordingly break the gradient and divergence into *radial* components and *horizontal* components, so that the radial component of the gradient is

$$\nabla_{\rm r} = \vec{e}_r \frac{\partial}{\partial r} \tag{3.18}$$

while the horizontal component is

$$\nabla_{\rm h} = \vec{e}_\theta \frac{1}{r}\frac{\partial}{\partial \theta} + \vec{e}_\phi \frac{1}{r \sin\theta}\frac{\partial}{\partial \phi} \tag{3.19}$$

Similarly, the Laplacian operator can be broken into radial and horizontal components, where the full operator is the sum of those two components, $\nabla^2 = \nabla_{\rm h}^2 + \nabla_{\rm r}^2$, so that

$$\nabla_{\rm h}^2 = \frac{\mathbf{L}^2}{r^2} = \frac{1}{r^2 \sin\theta}\frac{\partial}{\partial\theta}\left(\sin\theta\frac{\partial}{\partial\theta}\right) + \frac{1}{r^2\sin^2\theta}\frac{\partial^2}{\partial\theta^2} \tag{3.20}$$

Here the $\mathbf{L}^2$ represents an angular momentum operator, which some authors use rather than $\nabla_{\rm h}^2$. A very useful application of this for our purposes is that it allows us to write the eigenvalue equation:

$$L^2 Y_{lm}(\theta, \phi) = -r^2\left[\frac{1}{r^2\sin\theta}\frac{\partial}{\partial\theta}\left(\sin\theta\frac{\partial Y_{lm}}{\partial\theta}\right) + \frac{1}{r^2\sin^2\theta}\frac{\partial^2 Y_{lm}}{\partial\theta^2}\right] = l(l+1)Y_{lm}(\theta, \phi) \tag{3.21}$$

The quantity in brackets above is $\nabla_{\rm h}^2$, and we usefully note that operation on the spherical harmonic by this operator is the equivalent of multiplication by $\frac{-l(l+1)}{r^2}$.

We also have the radial component of the Laplacian,

$$\nabla_r^2 = \frac{1}{r^2}\frac{\partial}{\partial r}\left(r^2\frac{\partial}{\partial r}\right) \tag{3.22}$$

3.2 Oscillations

Consider a one-dimensional string with some linear density λ under tension. If we look at a short piece Δx of a longer string, then we can label the tension at one end of the short segment as T_1 and at the other as T_2, and similarly for the angles between the string and the horizontal θ_1, θ_2. Since the string is allowed to move only vertically, the horizontal forces on the piece must sum to zero, so in the horizontal direction we can write the force equation as

$$
\begin{aligned}
F_1 + F_2 &= ma_y \\
T \sin \theta_1 - T \sin \theta_2 &= mdx\frac{\partial^2 y}{\partial x^2} \\
-T\left(\frac{dy}{dx}\right)_1 + T\left(\frac{dy}{dx}\right)_2 &= \lambda dx\frac{\partial^2 y}{\partial x^2} \\
T\left[\left(\frac{dy}{dx}\right)_2 - \left(\frac{dy}{dx}\right)_1\right] &= \lambda dx\frac{\partial^2 y}{\partial x^2}
\end{aligned}
\tag{3.23}
$$

Dividing through by Δx and taking the limit as $\Delta x \to 0$ gives

$$
\frac{\partial^2 y}{\partial x^2} = \frac{\lambda}{T}\frac{\partial^2 y}{\partial x^2}
\tag{3.24}
$$

We recognize this as the one-dimensional wave equation, with propagation velocity

$$
c = \sqrt{\frac{\lambda}{T}}
\tag{3.25}
$$

Applying boundary conditions that the string remains fixed at both ends, so $y = 0$ at $x = 0$ and $x = L$, gives us solutions of the form

$$
y_n = A \sin k_n x
\tag{3.26}
$$

where n is an integer such that

$$
\begin{aligned}
k_n &= \frac{n\pi}{L} \\
\omega_n &= k_n c
\end{aligned}
\tag{3.27}
$$

There is a family of oscillatory solutions (eigenfunctions), each with a corresponding eigenfrequency. It's important to note that the velocity of the wave is determined by the physical parameters (density and tension) of the string, and that the structure of the wave (number of nodes) is determined by both the parameters of the string and the particular eigenfunction under consideration.

In the case of an oscillating star, we have a three-dimensional situation (r, θ, ϕ) rather than the one-dimensional string, but again we will have oscillatory solutions whose eigenfunctions and eigenfrequencies will tell us about physical conditions within the star.

3.3 Conservation Laws

In general, we will discuss the motion of a parcel of gas against the background of an equilibrium stellar model, and we will consider the properties of the gas to constitute a field, which can be either scalar or vector. There are two equivalent descriptions which can be used, and we will at times make use of both. First is the *Eulerian* approach, where the observer is located at a fixed position $\xi_0 = (r_0, \theta_0, \phi_0)$ and the parcel moves relative to that position. The second approach is the *Lagrangian* description, where the observer follows along with the moving gas parcel, so that its velocity relative to that observer is zero. The two approaches are connected, so that if we are describing a scalar field f we can write

$$\frac{df}{dt} = \frac{\partial f}{dt} + \nabla f \cdot \frac{d\boldsymbol{\xi}}{dt} = \frac{\partial f}{dt} + v \cdot \nabla \boldsymbol{\xi} \tag{3.28}$$

Extension to the case of a vector field is straightforward

$$\frac{d\mathbf{f}}{dt} = \frac{\partial \mathbf{f}}{dt} + v \cdot \nabla \boldsymbol{\xi} \tag{3.29}$$

3.3.1 Mass conservation

Mass conservation, sometimes expressed as *continuity*, simply states that the mass contained in a given volume element can change only if either a source or sink of mass is present (and therefore $\rho \neq$ constant), or if there is a net flow of mass either into or out through the boundaries of that volume element. Mathematically, we write

$$\frac{\partial \rho}{\partial t} + \rho(\nabla \cdot v) = \frac{\partial \rho}{\partial t} + (\nabla \cdot \rho v) = 0 \tag{3.30}$$

3.3.2 Linear Momentum Conservation

This can be alternatively expressed as the equation of motion for the parcel, but either way we write

$$\rho\frac{dv}{dt} = -\nabla P + \rho\mathbf{g} + \mathbf{F}_{\text{other}} \tag{3.31}$$

Here P is pressure and $\mathbf{g}$ the local gravity vector. We include $\mathbf{F}_{\text{other}}$ to represent additional forces that could be present, such as centripetal, Coriolis, magnetic, etc., though we will not consider that term going forward. Gravity $\mathbf{g}$ can be derived from a potential Φ

$$\mathbf{g} = -\nabla \Phi \tag{3.32}$$

where

$$\nabla^2\Phi = 4\pi G\rho \tag{3.33}$$

3.3.3 Energy Conservation

Here we write a thermodynamic expression stating that the change in the internal energy of the gas parcel q is equal to the heat supplied to it E and the work done by the parcel on its environment through expansion or contraction.

$$\frac{dq}{dt} = \frac{dE}{dt} + P\frac{dV}{dt} \tag{3.34}$$

We can use the continuity equation to rewrite this expression as

$$\frac{dq}{dt} = \frac{dE}{dt} + \frac{P}{\rho^2}\frac{d\rho}{dt} \tag{3.35}$$

If we assume the system is adiabatic, then we can further relate P and ρ through

$$P = P_0\left(\frac{\rho}{\rho_0}\right)^{\gamma} \tag{3.36}$$

Is adiabaticity a good assumption? It depends on the timescale of the oscillations compared to the timescale for heat transport across a length comparable to the physical scale of the oscillations. In general, this is a good assumption for most stars until we are very near the photosphere.

3.4 Important Frequencies

3.4.1 Buoyancy Frequency

Let's revisit the convective instability criterion from Section 1.5. Start by considering a small volume element or parcel of gas within the star with pressure and density P and ρ, that is in hydrostatic equilibrium and consider what happens if that element is displaced by some small distance δr. In general, the environment surrounding the element will now be different, with some new pressure and density P' and ρ', though we can take the differences from the initial position to be small. What happens to the parcel of gas? If we label the parcel's new thermodynamic variables with an asterisk and require pressure balance with its surroundings, then the pressure of the parcel is $P^* = P'$. Since the displacement δr is small, we assume that things are happening fast, so that there's no heat flow across the boundary of the parcel, or not enough to matter —the system is *adiabatic*, so the temperature within the parcel doesn't change either.

However, the density *will* change. Since the system is adiabatic, we know from Equation (1.39) that

$$p \sim \rho^{\gamma} \tag{3.37}$$

so

$$\rho^* = \rho\left(\frac{P'}{P}\right)^{1/\gamma} \tag{3.38}$$

we can relate the pressure at the new location to that at the old location through

$$P' = P + \frac{dP}{dr}\delta r \tag{3.39}$$

where the environment has some radial pressure gradient dP/dr. Similarly, for the density we have

$$\rho' = \rho + \frac{d\rho}{dr}\delta r \tag{3.40}$$

We can use the pressure expression from Equation (3.39) to get

$$\begin{aligned} \rho^* &= \rho\left(\frac{P'}{P}\right)^{1/\gamma} \\ &= \rho\left(\frac{P + \frac{dP}{dr}\delta r}{P}\right)^{1/\gamma} \\ &= \rho\left(1 + \frac{1}{P}\frac{dP}{dr}\delta r\right)^{1/\gamma} \end{aligned} \tag{3.41}$$

Since the second term inside the parentheses is small, we Taylor-expand and retain the first term to get

$$\rho^* = \rho\left(1 + \frac{1}{\gamma P}\frac{dP}{dr}\delta r\right) \tag{3.42}$$

We can now use Equation (3.40) to write the change in density $(\rho^* - \rho')$ as

$$\rho^* - \rho' = \left(\frac{1}{\gamma P}\frac{dP}{dr} - \frac{d\rho}{dr}\right)\delta r \tag{3.43}$$

If the term in parentheses is positive, then the parcel will feel a restoring buoyant force, and the system will be stable; if not, small displacements will tend to grow into large displacements and we will have the onset of convective motions. This is one version of the Schwarzschild stability criterion. A helpful alternative way to write the criterion makes use of the fact that

$$\frac{d \log \rho}{dr} = \frac{1}{\rho}\frac{d\rho}{dr} \tag{3.44}$$

and

$$\frac{d \log P}{dr} = \frac{1}{P}\frac{dP}{dr} \tag{3.45}$$

to render it as

$$\rho^* - \rho' = \rho\left(\frac{1}{\gamma}\frac{d \log P}{dr} - \frac{d \log \rho}{dr}\right)\delta r \tag{3.46}$$

The actual net buoyant force on the parcel is just equal to

$$F_{\text{net}} = g(\rho^* - \rho') \tag{3.47}$$

which allows us to write an equation of motion for the parcel

$$
\begin{aligned}
\frac{d^2(\delta r)}{dt^2} &= -\, g(\rho' - \rho^*) \\
&= -\, g\left(\frac{1}{\gamma}\frac{d\log P}{dr} - \frac{d\log\rho}{dr}\right)\delta r
\end{aligned}
\tag{3.48}
$$

We recognize this as the equation for one-dimensional simple harmonic motion at a frequency N given by

$$
N^2 = \left(\frac{1}{\gamma}\frac{d\log P}{dr} - \frac{d\log\rho}{dr}\right)
\tag{3.49}
$$

as long as $N^2 < 1$, corresponding to the Schwarzschild stability criterion. In this case, the parcel will execute simple harmonic motion about its equilibrium point at frequency N, where N is known as the *buoyancy frequency* or the *Brunt–Väisälä frequency*.

It is sometimes useful to write the Brunt–Väisälä frequency in a different form:

$$
\begin{aligned}
N^2 &= \left(\frac{1}{\gamma}\frac{d\log P}{dr} - \frac{d\log\rho}{dr}\right) \\
&= g\left(\frac{1}{\gamma P}\frac{dP}{dr} - \frac{1}{\rho}\frac{d\rho}{dr}\right)
\end{aligned}
\tag{3.50}
$$

Using

$$
c^2 = \frac{\gamma P}{\rho}
\tag{3.51}
$$

this reduces to

$$
N^2 = g\left(\frac{1}{c^2\rho}\frac{dP}{dr} + \frac{1}{H_\rho}\right)
\tag{3.52}
$$

where H_ρ is the density scale height. Further making use of the identity

$$
\frac{dP}{dr} = -\rho g
\tag{3.53}
$$

we get

$$
N^2 = g\left(\frac{-g}{c^2} + \frac{1}{H_\rho}\right) N^2 = \frac{-g^2}{c^2} + \frac{g}{H_\rho}
\tag{3.54}
$$

3.4.2 Acoustic Cut-Off Frequency

In Section 3.2 we briefly considered waves on a string, deriving the eigenfrequencies to be

$$\omega_n = k_n c = \frac{cn\pi}{L} \tag{3.55}$$

implying that the lowest sustainable frequency ω_0 is of order

$$\omega_0 = \frac{c\pi}{L} \tag{3.56}$$

We anticipate similar behavior in a gas, though in this case the restoring force is pressure rather than tension, so rather than the length of the string L we expect the boundary conditions to be set by the pressure scale height $H = kT/\mu m_{\rm H} g$, which would imply $\omega_0 \sim c/H$. A more exacting derivation gives the *acoustic cut-off frequency* a_c as

$$a_c = \frac{c}{H} \tag{3.57}$$

Waves with frequencies below this cutoff are *evanescent*, meaning their amplitudes decay exponentially, while those with frequencies above it can propagate. In practice, the acoustic cut-off frequency becomes important in or near the surface layers of the star.

3.5 Types of Waves

The two most common forms of waves encountered in the study of stellar oscillations are acoustic waves and internal gravity waves, so we begin by briefly summarizing their physics and properties. We will start from the equilibrium case outlined in Chapter 1, and assume local adiabaticity.

3.5.1 Acoustic Waves

For acoustic waves, we further assume that the local structure, including the gravitational potential, is constant, or at least varies slowly relative to the appropriate time constant for the wave. In this case, we can write the equation of motion for a small perturbation to the equilibrium state as

$$\rho_0 \frac{\partial^2 \delta\xi}{\partial t^2} = -\nabla P \tag{3.58}$$

Taking the divergence of both sides gives

$$\rho_0 \nabla \cdot \left(\frac{\partial^2 \delta\xi}{\partial t^2} \right) = -\nabla \cdot (\nabla P) \tag{3.59}$$

or

$$\rho_0 \frac{\partial^2}{\partial t^2}(\nabla \cdot \delta\xi) = - \nabla^2 P \tag{3.60}$$

From continuity we have

$$\rho + \nabla \cdot (\rho_0 \delta\xi) = 0 \tag{3.61}$$

so

$$\nabla \cdot \delta\xi = -\frac{\rho}{\rho_0} \tag{3.62}$$

Substituting then gives

$$\frac{\partial^2}{\partial t^2}\rho = - \nabla^2 P \tag{3.63}$$

In addition, adiabaticity implies

$$\Gamma_1 = \gamma = \left(\frac{d \log P}{d \log \rho}\right)_S \tag{3.64}$$

or

$$\nabla^2 P = \frac{\gamma \rho_0}{P_0} \nabla^2 \rho \tag{3.65}$$

and

$$\frac{\partial^2 P}{\partial t^2} = \frac{\gamma \rho_0}{P_0} \nabla^2 \rho \tag{3.66}$$

which is a wave equation describing plane acoustic waves traveling with sound speed

$$c_0 = \sqrt{\frac{\gamma P_0}{\rho_0}} \tag{3.67}$$

3.5.2 Internal Gravity Waves

While instructive in the sense that it confirms that propagating sound waves are possible in a medium representative of a stellar interior, and an indication of the speed of propagation of those waves, for internal gravity waves, there is a pressure gradient across the layer, given in the usual way by

$$\frac{dP}{dP_0} = -gP \tag{3.68}$$

in regions where the gravitational potential is constant, or at least slowly varying. If we neglect the $\mathbf{F}_{\text{other}}$ term, the general equation of motion for a gas parcel is Equation (3.31)

$$\rho\frac{dv}{dt} = -\nabla P + \rho\mathbf{g} \tag{3.69}$$

We perturb the variables in the equation, such that $X \rightarrow X + X'$, where X' represents the perturbation, and simplify to get an equation of motion in terms of the perturbed quantities:

$$\rho'\left(\frac{\partial^2 \boldsymbol{\xi}}{\partial t^2}\right) = -\nabla P' + \rho\mathbf{g}' + \rho'\mathbf{g} \tag{3.70}$$

As is now usual, we can look at the ξ_r and ξ_h portions independently, and assume that all variables are periodic, varying as $\exp[i(\mathbf{k}\cdot\boldsymbol{\xi} - \omega t)]$. Making the substitution produces

$$\begin{aligned} \rho\omega^2\xi_r &= ik_r P' + \rho' g \\ \rho\omega^2\boldsymbol{\xi}_\mathbf{h} &= i\mathbf{k}_\mathrm{h} P' \end{aligned} \tag{3.71}$$

The continuity Equation (3.30) can be solved in the same way to get

$$\rho' + \rho i k_r \xi_r + \rho i(\mathbf{k}_\mathrm{h} \cdot \xi_\mathrm{h}) \tag{3.72}$$

Now apply $\mathbf{k}_\mathrm{h}\cdot$ to the second equation in Equation (3.71) to solve for the pressure perturbation

$$P' = \frac{\omega^2}{k_\mathrm{h}^2}(\rho' + ik_r\rho\xi_r) \tag{3.73}$$

We use this in the first equation in Equation (3.71), discard the smaller term in ρ', and end up with

$$\rho\omega^2\left(1 + \frac{k_r^2}{k_\mathrm{h}^2}\right)\xi_r = \rho' g \tag{3.74}$$

Some further tedious manipulations result in

$$\omega^2\left(1 + \frac{k_r^2}{k_\mathrm{h}^2}\right)\xi_r = N^2\xi_r \tag{3.75}$$

where N is the buoyancy frequency. We can easily solve this to write a dispersion relation,

$$\omega^2 = \frac{N^2}{1 + k_r^2/k_\mathrm{h}^2} \tag{3.76}$$

We anticipated the buoyancy frequency to show up in a discussion of gravity waves, and are not disappointed. In addition, note that the solutions to this equation of motion is oscillatory only when $N^2 > 0$, which we can also write as

$$N^2 = \frac{d\log\rho}{d\log P} > 0 \tag{3.77}$$

which we identify as the criterion for convective stability. Here the interpretation is slightly different: there is no oscillatory solution for gravity modes in regions where energy transport is by convection! In those regions, solutions to the equations of motion exhibit exponential decay, or are evanescent.[2]

3.6 The Stellar Oscillation Equations

Let's start by looking at perturbing the equations of motion, starting from Equation (3.31).

$$(\rho + \rho')\frac{d(v + v')}{dt} = - \nabla (P + P') + (\rho + \rho') \nabla (\Phi + \Phi') \tag{3.78}$$

Dividing through gives

$$\frac{d(v + v')}{dt} = \frac{- \nabla (P + P')}{(\rho + \rho')} + \nabla (\Phi + \Phi') \tag{3.79}$$

or

$$\frac{d(v + v')}{dt} = \frac{- \nabla (P + P')}{\rho(1 + \rho'/\rho)} + \nabla (\Phi + \Phi') \tag{3.80}$$

Taylor-expanding the denominator of the first term on the right allows us to write this as

$$\frac{dv}{dt} + \frac{dv'}{dt} \approx -\frac{1}{\rho} \nabla (P + P')(1 - \rho'/\rho) + \nabla (\Phi) + \nabla (\Phi') \tag{3.81}$$

Multiplying through, subtracting the original unperturbed expression from both sides, and discarding terms which are second-order in the perturbation, leaves

$$\rho\frac{\partial^2 \boldsymbol{\xi}}{\partial t^2} + \nabla P' - \frac{\rho'}{\rho} \nabla P + \rho \nabla \Phi' = 0 \tag{3.82}$$

As a reminder, here P and ρ represent pressure and density and Φ is the gravitational potential, which satisfies Poisson's equation

$$\nabla^2\Phi = -4\pi G\rho \tag{3.83}$$

Primed quantities here represent perturbed values.

We assume that both the position ξ of a parcel undergoing motion and the various physical parameters such as pressure and density can all be decomposed into horizontal, radial, and temporal components, so that

$$\xi = e^{i\omega t}\xi_{\rm r}(r) Y_{lm}(\theta, \phi) \tag{3.84}$$

[2] An analogy is found in quantum mechanical tunneling.

While we aren't going to prove this here, it's a reasonable assumption as the spherical harmonics represent a basis set in spherical polar coordinates, and the complex exponential $e^{i\omega t}$ represents oscillatory motion, which is what we generally anticipate.

Since the radial and horizontal components are separable, we can now examine the equation of motion in both independently. Working horizontally, we have

$$\rho\frac{\partial^2\xi_h}{\partial t^2} + \nabla_h \; P' - \frac{\rho'}{\rho}\nabla_h \; P + \rho\,\nabla_h \; \Phi' = 0 \tag{3.85}$$

Here the third term vanishes because the pressure gradient in the horizontal direction does too, because the equilibrium configuration of the star is spherically symmetric, leaving

$$\rho\frac{\partial^2\xi_h}{\partial t^2} + \nabla_h \; P' + \rho\,\nabla_h \; \Phi' = 0 \tag{3.86}$$

We now apply the horizontal divergence operator to both sides

$$\nabla_h \cdot \left(\rho\frac{\partial^2\xi_h}{\partial t^2} + \nabla_h \; P' + \rho\,\nabla_h \; \Phi'\right) = 0 \tag{3.87}$$

resulting in

$$\rho\frac{\partial^2}{\partial t^2}(\nabla_h \cdot \xi_h) + \nabla_h^2 \; P' + \rho\,\nabla_h^2 \; \Phi' = 0 \tag{3.88}$$

Now we can make use of the equation of continuity, Equation (3.30), to simplify this expression, by substituting for the $(\nabla_h \cdot \xi_h)$. Start with

$$\rho' + (\nabla_r + \nabla_h) \cdot (\rho\xi) = 0 \tag{3.89}$$

Rearranging

$$\rho' = -(\nabla_r \cdot \rho\xi_r) - (\nabla_h \cdot \rho\xi_h) \tag{3.90}$$

and making use of the ∇_r operator

$$\rho' = -\frac{1}{r^2}\frac{\partial}{\partial r}(\rho r^2\xi_r) - (\nabla_h \cdot \rho\xi_h) \tag{3.91}$$

which we can solve for $(\nabla_h \cdot \xi_h)$:

$$\nabla_h \cdot \xi_h = \frac{1}{\rho}\left[-\rho' - \frac{1}{r^2}\frac{\partial}{\partial r}(\rho r^2\xi_r)\right] \tag{3.92}$$

Now we can use this in our horizontal equation of motion expression (3.88) to get

$$-\frac{\partial^2}{\partial t^2}\left[\rho' + \frac{1}{r^2}\frac{\partial}{\partial r}(\rho r^2\xi_r)\right] + \nabla_h^2 \; P' + \rho\,\nabla_h^2 \; \Phi' = 0 \tag{3.93}$$

Next we can make use of our eigenvalue equation for $\mathbf{L}^2$ (Equation (3.21)), so that this becomes

$$-\omega^2\left[\rho' + \frac{1}{r^2}\frac{\partial}{\partial r}(\rho r^2 \xi_r)\right] - \frac{l(l+1)}{r^2}P' - \frac{l(l+1)}{r^2}\rho\Phi' = 0 \tag{3.94}$$

Our assumption of adiabaticity will allow us to eliminate ρ' from the expression.

Now let's look at the radial component of the equation of motion.

$$\frac{P' + R(dP/dr)}{P} = \Gamma_1 \frac{\rho' + R(d\rho/dr)}{\rho} \tag{3.95}$$

In the radial direction, we can write

$$\rho\frac{\partial^2 \xi_r}{\partial t^2} + \nabla_r\ P' - \frac{\rho'}{\rho}\nabla_r\ P + \rho\,\nabla_r\ \Phi' = 0 \tag{3.96}$$

which substituting for the radial component of the gradient ∇_r gives

$$\rho\frac{\partial^2 \xi_r}{\partial t^2} + \frac{\partial P'}{\partial r} - \frac{\rho'}{\rho}\frac{\partial P}{\partial r} + \rho\frac{\partial \Phi'}{\partial r} = 0 \tag{3.97}$$

We can further simplify by recognizing that

$$\frac{\partial P}{\partial r} = \rho g \tag{3.98}$$

so we have for the radial component of the equation of motion

$$\rho\frac{\partial^2 \xi_r}{\partial t^2} + \frac{\partial P'}{\partial r} - \rho' g + \rho\frac{\partial \Phi'}{\partial r} = 0 \tag{3.99}$$

and doing our usual trick with the time derivative turns this into

$$-\omega^2 \rho \xi_r + \frac{\partial P'}{\partial r} - \rho' g + \rho\frac{\partial \Phi'}{\partial r} = 0 \tag{3.100}$$

In addition, we have the Poisson equation for the gravitational potential

$$\nabla^2 \Phi' = -4\pi G \rho' \tag{3.101}$$

which we can rewrite as

$$\left(\nabla_r^2 + \nabla_h^2\right)\Phi' = \frac{1}{r^2}\frac{d}{dr}\left(r^2\frac{d\Phi'}{dr}\right) - \frac{l(l+1)}{r^2}\Phi' = -4\pi G\rho' \tag{3.102}$$

At this point we will assume the oscillations to be adiabatic, which implies that they occur rapidly enough that we can ignore energy transport into or out of a gas parcel. Intuitively, we can expect that this assumption might be challenged near the surface, but as long as we stay deep enough inside the stellar envelope we can take

$$\frac{\delta P}{P} \sim \frac{\delta \rho}{\rho} = \gamma \frac{\delta \rho}{\rho} \tag{3.103}$$

where γ is the adiabatic exponent,[3] defined by

$$\gamma = \left(\frac{d \log P}{d \log \rho} \right)_S \tag{3.104}$$

and related to the sound speed as

$$c^2 = \gamma \frac{P}{\rho} \tag{3.105}$$

We now have four equations for four unknowns. With a little further effort, we can rewrite these in terms of physically meaningful parameters. In particular, we will use the Brunt–Väisälä (or buoyancy) frequency N,

$$N^2 = g \left(\frac{1}{\Gamma_1 P} \frac{dP}{dr} - \frac{1}{\rho} \frac{d\rho}{dr} \right) \tag{3.106}$$

and define the *characteristic acoustic frequency*[4] S_l such that

$$S_l^2 = \frac{l(l+1)c^2}{r^2} = k_{\mathrm{h}}^2 c^2 \tag{3.107}$$

Start from

$$\frac{P' + \xi_{\mathrm{r}}(dP/dr)}{P} = \gamma \frac{\rho' + \xi_{\mathrm{r}}(d\rho/dr)}{\rho} \tag{3.108}$$

Rearranging gives

$$\frac{\gamma P}{\rho}(\rho' + \xi_{\mathrm{r}} dP/dr) = P' + \xi_{\mathrm{r}} dP/dr \tag{3.109}$$

$$\rho' = \frac{\rho P'}{\gamma P} + \rho \xi_{\mathrm{r}} \left[\frac{1}{\gamma P} \frac{dP}{dr} - \frac{1}{\rho} \frac{d\rho}{dr} \right] \tag{3.110}$$

This result can now be used to substitute for ρ' in Equation (3.100), giving

$$\frac{d\xi}{dr} = -\left(\frac{2}{r} + \frac{1}{\gamma P} \frac{dP}{dr} \right) \xi_{\mathrm{r}} - \frac{1}{\rho c^2} \left(\frac{S_l^2}{\omega^2} - 1 \right) P' - \frac{l(l+1)}{\omega^2 r^2} \Phi' \tag{3.111}$$

Here we've made use of the Lamb frequency S_l.

[3] Formally, it's the *first* adiabatic exponent, Γ_1.

[4] Also called the *Lamb frequency*.

We can similarly simplify the radial component of the equation of motion, recognizing that the quantity in parentheses in Equation (3.110) is just the buoyancy frequency N

$$\rho\xi_{\mathrm{r}}\left[\frac{1}{\Gamma_1 P}\frac{dP}{dr} - \frac{1}{\rho}\frac{d\rho}{dr}\right] = \frac{\rho\xi_{\mathrm{r}}N^2}{g} \tag{3.112}$$

so making use of this causes that equation to reduce to

$$\frac{dP'}{dr} = \rho(\omega^2 - N^2)\xi_{\mathrm{r}} + \frac{1}{\Gamma_1 P}\frac{dP}{dr}P' + \rho\frac{d\Phi'}{dr} \tag{3.113}$$

Finally, substituting for ρ' into the Poisson equation,

$$\frac{1}{r^2}\frac{d}{dr}\left(r^2\frac{d\Phi'}{dr}\right) - \frac{l(l+1)}{r^2}\Phi' = -4\pi G\rho', \tag{3.114}$$

gives

$$\frac{1}{r^2}\frac{d}{dr}\left(r^2\frac{d\Phi'}{dr}\right) = \frac{l(l+1)}{r^2}\Phi' - 4\pi G\left(\frac{P'}{c^2} + \frac{\rho\xi_{\mathrm{r}}N^2}{g}\right) \tag{3.115}$$

Together these three fourth-order equations are a complete set for the four variables ξ_{r}, ρ', Φ', and $\frac{d\Phi'}{dr}$.

3.6.1 Cowling Approximation

The equations shown above are amenable to numerical solution, but it's extremely helpful to write them in a way which is conducive to our physical intuition. We can move along that path by assuming that the scale length of the oscillations is small compared to the stellar scale length, the radius. This will be true for waves of high order. This is the *Cowling approximation*, and we adopt it by taking $\Phi' = 0$, so we neglect changes in local gravity. This immediately reduces the equations of motion to

$$\frac{d\xi}{dr} = -\left(\frac{2}{r} + \frac{1}{\Gamma_1 P}\frac{dP}{dr}\right)\xi_{\mathrm{r}} + \frac{1}{\rho c^2}\left(1 - \frac{S_l^2}{\omega^2}\right)P' \tag{3.116}$$

$$\frac{dP'}{dr} = \rho(\omega^2 - N^2)\xi_{\mathrm{r}} + \frac{1}{\Gamma_1 P}\frac{dP}{dr}P' \tag{3.117}$$

We can (somewhat brutally!) simplify further by recognizing that, particularly for higher-frequency oscillations, the terms containing derivatives of equilibrium quantities ($\frac{dP}{dr}$, in particular) are very much smaller than the others, so we neglect these as well, giving

$$\frac{d\xi}{dr} = \frac{1}{\rho c^2}\left(1 - \frac{S_l^2}{\omega^2}\right)P' \tag{3.118}$$

$$\frac{dP'}{dr} = \rho(\omega^2 - N^2)\xi_r \tag{3.119}$$

Finally, taking the radial derivative of the expression for ξ_r and using Equation (3.117) to substitute for the resulting dP'/dr allows us to combine the two into a single second-order differential equation

$$\frac{d^2\xi_r}{dr^2} = \frac{\omega^2}{c^2}\left(1 - \frac{S_l^2}{\omega^2}\right)\left(1 - \frac{N^2}{\omega^2}\right)\xi_r \tag{3.120}$$

This is immediately recognizable as the equation for oscillatory motion, of the form

$$\frac{d^2\xi_r}{dr^2} = K^2\xi_r \tag{3.121}$$

where

$$K(r) = \frac{\omega^2}{c^2}\left(1 - \frac{S_l^2}{\omega^2}\right)\left(1 - \frac{N^2}{\omega^2}\right) \tag{3.122}$$

For oscillations to occur, we require $K(r) < 1$. Rewriting the expression for $K(r)$ makes the conditions for this to occur more obvious:

$$K(r) = \frac{1}{\omega^2 c^2}\left(\omega^2 - S_l^2\right)(\omega^2 - N^2) \tag{3.123}$$

Sound waves can be described using a dispersion relation of the general form

$$\omega^2 = c^2\,|k|^2 \tag{3.124}$$

Inspection shows that in this case we can identify $K(r)$ with the wavenumber k, and we can break k into radial and horizontal components, where

$$k^2 = k_r^2 + k_h^2 \tag{3.125}$$

By inspection, $K(r) < 0$ when either

$$\omega^2 > S_l^2 \textbf{ and } \omega^2 > N^2 \tag{3.126}$$

of

$$\omega^2 < S_l^2 \textbf{ and } \omega^2 < N^2 \tag{3.127}$$

The first case corresponds to pressure, or p-modes, which have pressure as their restoring force and occupy the high-frequency allowed space, and gravity, or g-modes, which have gravity as their restoring force and occupy the low-frequency allowed space. Figure 3.1 illustrates the breakdown for a typical solar-like star.

3.6.2 Dispersion Relations

It's informative to look at the special cases when ω is large and when ω is small. Let's start with the former first, and consider the high-frequency case, where $\omega^2 \gg N^2$. In the case of the Sun, these are relatively high-frequency waves (the so-called "5 minute

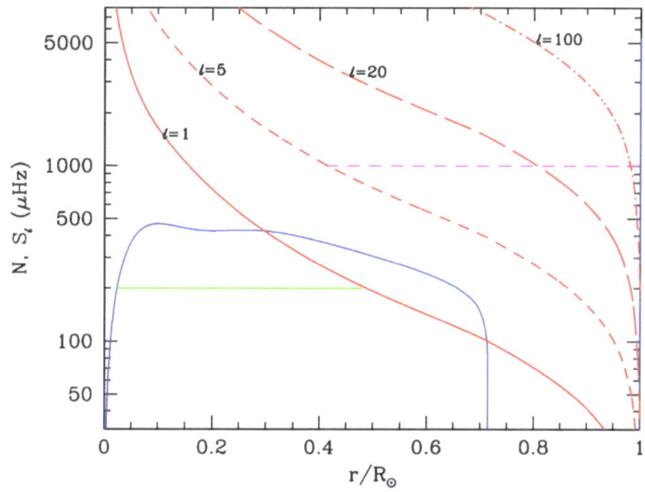

Figure 3.1. The propagation diagram for a standard solar model. The blue line is the Brunt–Väisälä (buoyancy) frequency, the red lines are the Lamb (characteristic) frequencies as a function of depth for different degrees l. The green solid horizontal line shows the region where a $200\mu Hz$ g-mode can propagate. The pink dashed horizontal line shows where a $1000\mu Hz$, $l = 5$ p-mode can propagate. Note that low-degree p-modes can propagate only near the surface, while g-modes are confined to a cavity near the stellar core. Reprinted with permission from Basu (2016).

oscillations") trapped in a cavity near the surface. What are the boundaries of the cavity? The inner boundary occurs when $S_l = \omega$, which occurs when

$$S_l^2 = \frac{l(l+1)c^2}{r^2} = \omega \tag{3.128}$$

or

$$r = \sqrt{\frac{l(l+1)c^2}{\omega}} \tag{3.129}$$

The outer turning point is approximately the stellar radius, though this is only an approximation.

In that case, we can approximate the dispersion relation as

$$K(r) = \frac{\omega^2}{c^2}\left(1 - \frac{S_l^2}{\omega^2}\right)\left(1 - \frac{N^2}{\omega^2}\right) = \frac{\omega^2}{c^2}\left(1 - \frac{S_l^2}{\omega^2}\right) = \frac{\omega^2 - S_l^2}{c^2} \tag{3.130}$$

Just as we broke the operator ∇^2 into radial and horizontal components so $\nabla^2 = \nabla_\mathrm{r}^2 + \nabla_\mathrm{h}^2$, we can break the wavevector k into components so that $k^2 = k_\mathrm{r}^2 + k_\mathrm{h}^2$. Examination of our expression for S_l^2 at the turning point allows us to recognize k_h^2

$$k_\mathrm{h}^2 = \frac{l(l+1)}{r^2} \tag{3.131}$$

so we have

$$c^2K = \omega^2 - k_{\rm h}^2c^2 \tag{3.132}$$

Rearranging we have

$$\omega^2 = c^2\left(K + k_{\rm h}^2\right) \tag{3.133}$$

and we can identify K as $k_{\rm r}^2$.

$$K = k_{\rm r}^2 = \frac{\omega^2 - S_l^2}{c^2} \tag{3.134}$$

The wavenumber k for p-modes increases with frequency ω, so that higher order p-modes have higher frequencies.

For g-modes, we examine the low-frequency case, where $\omega^2 \gg S_l^2$. Here we can approximate K as

$$K(r) = \frac{\omega^2}{c^2}\left(1 - \frac{S_l^2}{\omega^2}\right)\left(1 - \frac{N^2}{\omega^2}\right) = \frac{S_l^2}{c^2}\left(\frac{N^2}{\omega^2} - 1\right) = \frac{l(l+1)}{r^2}\left(\frac{N^2}{\omega^2} - 1\right) \tag{3.135}$$

where in the last step we've made use of Equation (3.128). From this formulation, we can see that k increases with decreasing ω, so higher order g-modes have *lower* frequencies, or longer wavelengths.

Alternatively, we can start from

$$k_{\rm r}^2 = \frac{S_l^2}{\omega^2c^2}(N^2 - \omega^2) \tag{3.136}$$

and make use of the identity $S_L^2 = c^2k_{\rm h}^2$, so that

$$\omega^2\left(k_{\rm r}^2 + k_{\rm h}^2\right) = k_{\rm h}^2N^2 \tag{3.137}$$

$$\omega^2 = \frac{k_{\rm h}^2N^2}{k^2} = N^2\cos\theta \tag{3.138}$$

where we define θ as the angle between the wavevector and the local horizontal. Gravity modes primarily propagate horizontally.

We can illustrate the situation inside stars by use of a *propagation diagram*, as shown in Figure 3.1 for a star like the Sun. The diagram shows that, in general for solar-like stars, p-modes are confined to regions in the envelope and g-modes to regions near the core of the star, in bounded regions which effectively represent independent oscillation cavities within the star. However, at frequencies below about 500μHz, solutions exist in both cavities, which have g-mode character near the core and p-mode character in surface regions. These are the *mixed modes*.

In red giant stars, the propagation diagram is significantly more complex (Figure 3.2). As the g-mode and p-mode cavities draw close to one another, the evanescent region

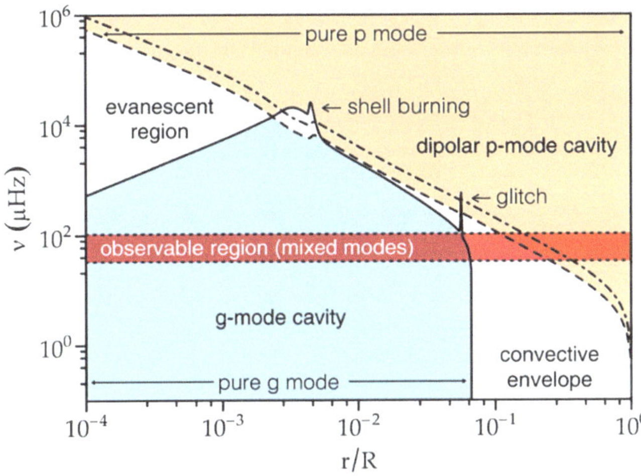

Figure 3.2. The significantly more complex propagation diagram for a red giant star. The "observable region" denotes the region around ν_{max}, while the dashed and dot-dashed curves show the characteristic (Lamb) frequencies for $l = 1$ and $l = 2$ as a function of depth. The p-mode and g-mode cavities bounded by the buoyancy and Lamb frequencies (respectively) are labeled, as are the evanescent regions in which periodic solutions do not exist. The so-called "glitch" visible in the buoyancy frequency is caused by the near-discontinuity in the hydrogen abundance resulting from the penetration of the convective envelope during first dredge-up into a region where the composition has been modified by nuclear reactions. Reprinted with permission from Christensen-Dalsgaard et al. (2020).

between them is no longer deep enough to prevent "tunneling" of modes between the cavities, giving rise to *mixed modes*, which combine both p- and g-mode character and whose frequencies are determined by physical conditions in both cavities. In addition, "glitches" in the buoyancy frequency curve become visible (see discussion in Section 8.4).

3.6.3 Asymptotic Relations

We can estimate the frequencies corresponding to the normal modes by applying a Bohr quantization criterion; essentially, that an integer number of wavelengths fit across the oscillation cavity. In general, this implies

$$\int_{r_1}^{r_2} k_r \, dr = \pi(n + \alpha) \tag{3.139}$$

Here k_r is the radial wavenumber, r_1 and r_2 are the boundaries of the resonant cavity (determined by locations where $k_r = 0$), n is the radial order of the oscillation, and α represents a phase shift. α is determined by the conditions near the boundaries of the cavity, where the $\pi/2$ phase shift one might anticipate from simple reflection is complicated by the fact that conditions near the boundary don't change instantaneously.

3.6.3.1 p Modes

For p-modes, we can write

$$\int_{r_1}^{R} \sqrt{\frac{\omega^2}{c^2} - \frac{L^2}{r^2}}\, dr = \pi(n + \alpha) \tag{3.140}$$

For small $L^2 = l(l + 1)$, we can simplify this to

$$\int_{r_1}^{R} \sqrt{\frac{\omega^2}{c^2}}\, dr = \pi(n + \alpha) \tag{3.141}$$

For small degree modes, the inner turning point is close to the center of the star, so we take $r_1 = 0$, and as before adopt $r_2 = R$, to get

$$\omega \simeq \frac{\pi(n + \alpha)}{\displaystyle\int_0^R \frac{dr}{c}} \tag{3.142}$$

Observationally, we measure $\nu = \omega/2\pi$, so rewriting

$$\nu \simeq \frac{(n + \alpha)}{2\displaystyle\int_0^R \frac{dr}{c}} = (n + \alpha)\Delta\nu \tag{3.143}$$

where

$$\Delta\nu = \left[2\int_0^R \frac{dr}{c}\right]^{-1} \tag{3.144}$$

is the so-called *large frequency separation*, representing the inverse of the sound travel time across the centerline of the star. The key observational insight here is that we anticipate that oscillation frequencies representing consecutive radial orders n and $n + 1$ are evenly spaced.

The second term within the radical of Equation (3.140) includes $L^2 = l(l + 1)$, implying that a more accurate version of the expression for p-mode frequencies would have a dependence not only on radial order n but also on degree l. In fact a higher-order treatment modifies the expression for p-mode frequencies to read

$$\nu = \left(n + \frac{l}{2} + \alpha\right)\Delta\nu \tag{3.145}$$

We therefore anticipate that p-modes of the same order n but differing by $\Delta l = 1$ are also evenly spaced, and that it's appropriate to label frequencies by both n and l, so that $\nu_{nl} = \nu_{n+1,l-2}$.

Continuing the analysis to even higher order gives

$$\nu = \left(n + \frac{l}{2} + \frac{1}{4} + \alpha\right)\Delta\nu - (AL^2 - \delta)\frac{\Delta\nu^2}{\nu_{nl}} \tag{3.146}$$

where

$$A = \frac{1}{4\pi^2\Delta\nu}\left[\frac{c}{R} - \int_0^R \frac{dc}{dr}\frac{dr}{r}\right] \tag{3.147}$$

This expansion shows that there are small departures from the regularity of $\Delta\nu$. We can define the *small separation* $\delta\nu$ such that

$$\delta\nu_{nl} = \nu_{nl} - \nu_{n-1,l+2} \approx -(4l+6)\frac{\Delta\nu}{4\pi^2\nu_{nl}}\int_0^R \frac{dc}{dr}\frac{dr}{r} = (4l+6)D_0 \tag{3.148}$$

Examination of the integral shows that it is sensitive to the sound speed gradient $\frac{dc}{dr}$ weighted inversely by the radial coordinate, meaning that it is most affected by conditions near the core. In particular, the composition near the core evolves over a star's main lifetime as the hydrogen fraction decreases due to nuclear fusion, so the small separation can be used to measure age on and near the main sequence.

3.6.3.2 g Modes

We can follow a similar prescription to understand g-mode frequencies in the asymptotic limit. Here recall that we have

$$k_r^2 = = \frac{L^2}{r^2\omega^2}(N^2 - \omega^2) \tag{3.149}$$

and we consider the regime where $N \gg \omega$, or $k_r \gg 0$. Physically this means that we are far from the boundaries of the acoustic cavity defined by the turning points for the wave, where $N = \omega$. Here, the expression for k_r simplifies, so

$$k_r \approx \frac{LN}{\omega r} \tag{3.150}$$

Quantizing that as we did for p-modes gives

$$\int_{r_1}^{r_2} \frac{LN}{\omega r} = \pi\left(n + \alpha_g\right) \tag{3.151}$$

As with p-modes there is a phase α_g set by the near-boundary conditions of the cavity; we've used a subscript g here as a reminder that the phase factors are different between the two cases! Solving for frequency ω yields

$$\omega = \frac{\frac{L}{\omega}\int_{r_1}^{r_2}\frac{N}{r}dr}{\pi\left(n + \alpha_g\right)} \tag{3.152}$$

Once again, n is the radial order, but here the modes are evenly spaced in ω^{-1} rather than ω, which translates into being evenly spaced in *period* rather than frequency. A second-order analysis gives

$$\omega = \frac{\frac{L}{\omega}\int_{r_1}^{r_2}\frac{N}{r}dr}{\pi\left(n + \frac{l}{2} + \alpha_g\right)} \tag{3.153}$$

The period spacing here is given by

$$\Delta\Pi_l = \frac{\Delta\Pi_0}{L} \tag{3.154}$$

where

$$\Delta\Pi_0 = \frac{2\pi^2}{\displaystyle\int_{r_1}^{r^2} \frac{Ndr}{r}} \tag{3.155}$$

3.7 The Physical Interpretation of Oscillation Modes

Stellar oscillations are resonant modes, just as are the oscillations on a string, and occur at specific frequencies. We treat them as small perturbations to the equilibrium description of the star, which is a good approximation for all but the largest-amplitude classical pulsators. Since deviations from spherical symmetry are presumed to be small perturbations, the problem is a separable one, and we end up with equations of the form of Equation (3.120), which are eigenvalue problems, and thus produce a spectrum of eigenfrequencies, each with a corresponding eigenmode. Note that in this case only the radial part of the solution depends on the physical structure of the star, while the horizontal θ, ϕ portion of the solution is then well-described by the spherical harmonics.

We then characterize each mode using three integers $\{nlm\}$, effectively spherical harmonic or quantum numbers. Figure 3.3 shows some examples of the spherical harmonic portion of each mode, labeled by $\{lm\}$. These numbers themselves are:

- n, the **radial order**, or just the **order**. This indicates the number of zeroes, or nodes, in the radial direction. For clarity, and by convention, $n > 0$ is taken to apply to p-modes and $n < 0$ to g-modes. The radial order is generally not directly observable.
- l, the **angular degree**, or just the **degree**, which indicates the total number of nodal lines (reversals) on the surface of the star. An oscillation mode with $l = 0$ is therefore one with no surface nodal lines, and is thus a purely radial mode. Non-radial modes have $l \geqslant 1$. Note that l is always a positive integer. For historical reasons, modes with $l = 1$ are sometimes known as dipole modes, $l = 2$ as quadrupole, and so on.
- m, the **azimuthal order**, gives the number of surface nodal lines which pass through the poles of the star. We have so far only considered non-rotating stars, but we will see that in a rotating star we can define the direction in which the oscillation propagates as either prograde (in the rotation direction) or retrograde. By convention, we take $m > 0$ as prograde. Note that it is possible to have $m = 0$, which corresponds to a purely axisymmetric, or *zonal* mode; in contrast, modes with $|m| > 0$ are *sectoral* modes.

Figure 3.4 repeats Figure 3.3, but showing only the nodal line boundaries, where the value of the solution changes sign, which can make it easier to visualize the behavior, while Figure 3.5 shows patterns for a mode of higher degree ($l = 6$), to more clearly show the difference between l and m.

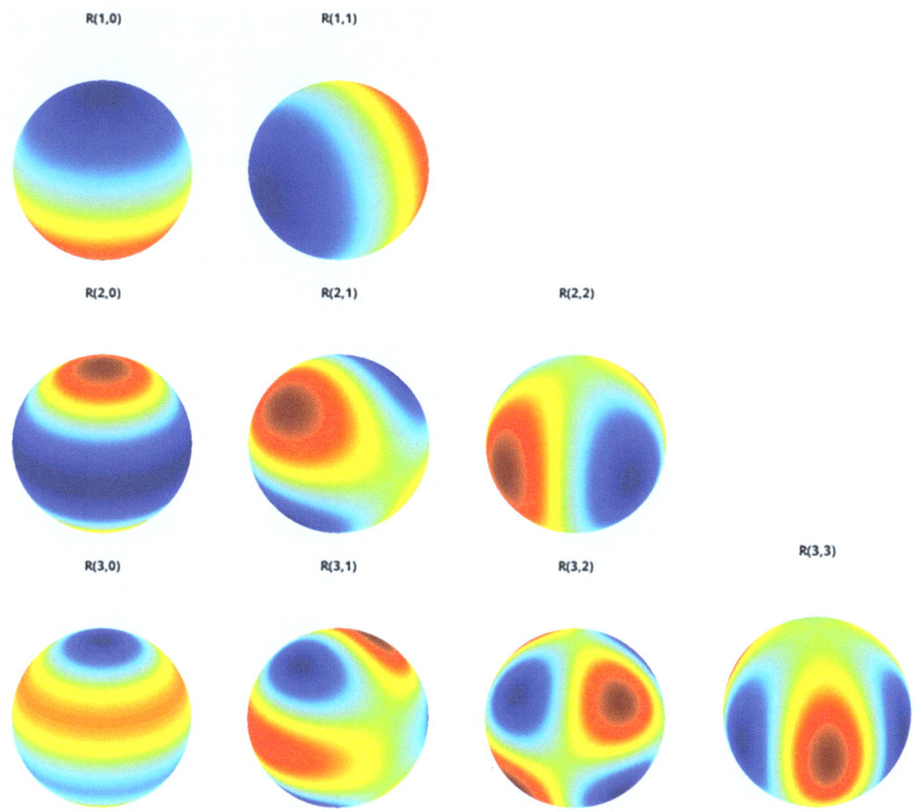

Figure 3.3. Spherical harmonics illustrated on the sphere, with color coding to local displacement. Each is labeled with $R(l, m)$ and only positive m are shown.

3.7.1 Ray Paths and Turning Points

In solar-like stars, p-modes are confined to the outer layers, with their resonant cavity bounded at the top by the stellar photosphere (a somewhat fuzzy boundary) and at the bottom by the so-called "internal turning point." Physically, the acoustic wave *reflects* off the top of the cavity due to the exponentially rapid decrease of pressure and density in the stellar atmosphere. The critical time τ_c for wave reflection is roughly the time for the wave traveling at the local sound speed c_s to travel a pressure scale height $H = kT/\mu m_H g$, so the critical frequency ω_c for reflection is

$$\omega_c \sim \frac{c_s}{H} \tag{3.156}$$

A more precise derivation can be found in Deubner & Gough (1984). The internal turning point can be understood in terms of diffraction: the sound speed increases with depth, and the direction of propagation then diverges from the radial path until the wave reverses direction.

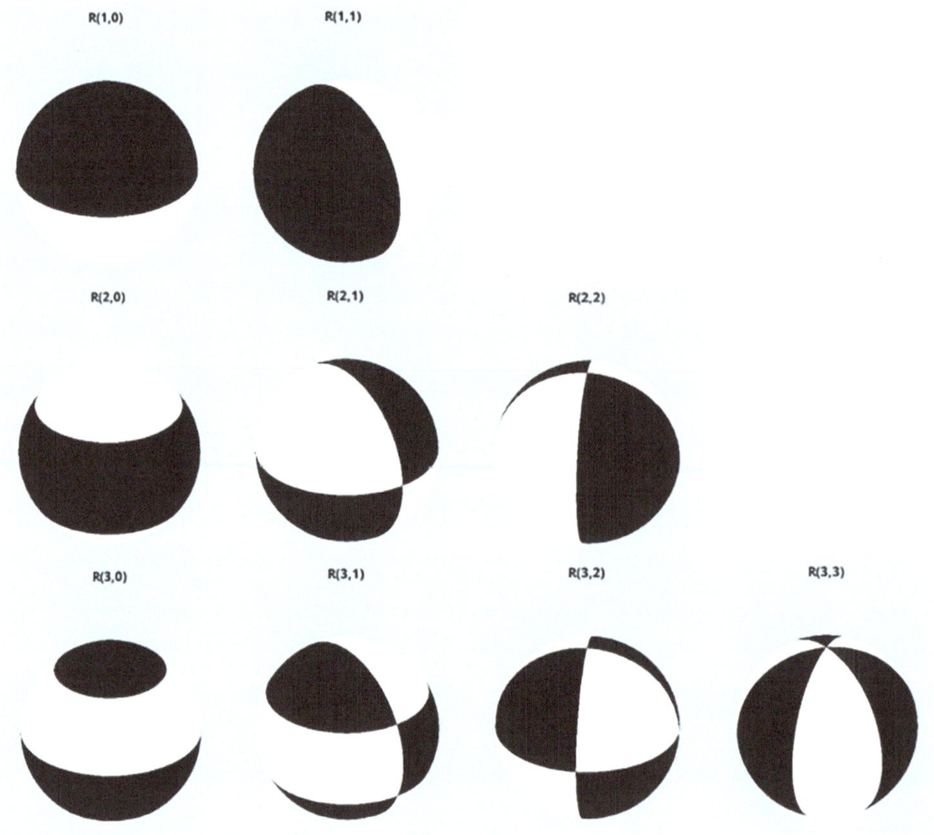

Figure 3.4. Same as Figure 3.3, but in black-and-wide coded to the local sign (±), to make the nodal lines more apparent.

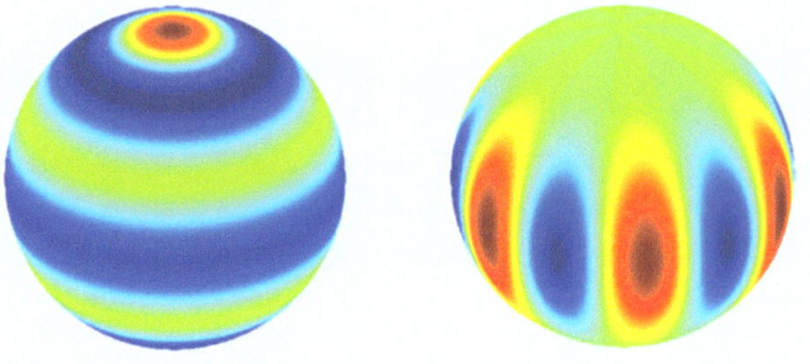

Figure 3.5. An example of a higher-order harmonic, showing n, $l = 6$, 0 on the left and n, $l = 6$, 6 on the right.

Figure 3.6 illustrates the paths between the inner and outer turning points for p-modes propagating within a solar-like star. Reflection at the outer turning point and refraction at the inner are both visible, as is the fact that the mode of higher

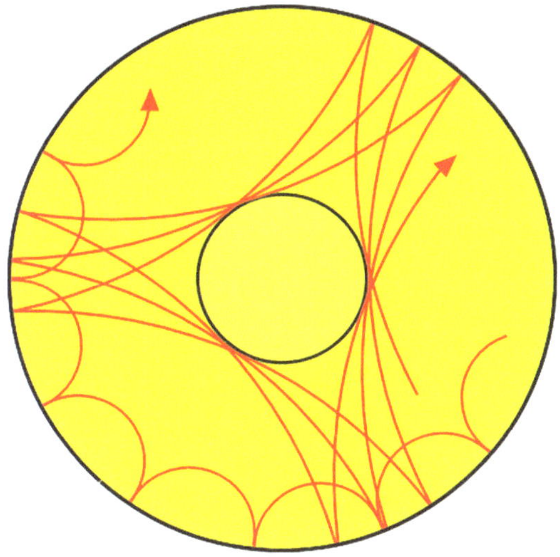

Figure 3.6. A ray diagram showing a low-degree mode, which penetrates deep into the stellar interior, and a high-degree mode, which samples only the near-surface regions. Effectively, the high-degree mode is only sensitive to conditions in the outer layers of the star.

degree l samples only outer portions of the star, while the lower-degree mode penetrates to the base of the convection zone in this case.

In helioseismic observations, the Sun is a resolved source, so the observable limit on angular degree l is large (hundreds) and is potentially constrained by the spatial resolution of the observations. In contrast, stars other than the Sun are unresolved sources, so the signal we detect from high-degree modes is heavily diluted and typically only modes of $l \leqslant 3$ can be measured. However, in a happy coincidence, these are also the modes which penetrate the deepest into the star.

In contrast to the p-modes, g-modes in solar-like stars are confined to the interior of the star, since they cannot propagate in convective zones. In the other envelope, these waves are evanescent and their amplitude at the surface is expected to be extremely small in stars like the Sun; in fact, the detection of g-modes in the Sun itself, the star for which data are the best, has been reported on and off for years without yet being confirmed (Appourchaux et al. 2010). However, in high-mass stars with radiative envelopes, g-modes are frequently detected from observations of the surface.

3.8 Mode Inertia

An important characteristic of any oscillation mode is the mode inertia. We can write the kinetic energy of an oscillation mode as

$$E_{\mathrm{kin}} = \frac{1}{2}\int_V \rho|v|^2 dV \tag{3.157}$$

which is proportional to the mean square value of the velocity. To understand this in terms of quantities we already know, we break the mean square displacement into radial and horizontal components, $\delta r^2_{\rm rms}$ and $\delta h^2_{\rm rms}$. We can calculate the values of these as

$$\delta r^2_{\rm rms} = \frac{1}{\Pi}\int_0^\Pi dt\left[\frac{1}{4\pi}\mathscr{R}\left(\xi_{\rm r}(r)\,Y_l^m(\theta,\,\phi)e^{-i\omega t}\right)^2 d\Omega\right] = \frac{1}{2}|\xi_{\rm r}(r)|^2 \tag{3.158}$$

and similarly

$$\delta h^2_{\rm rms} = \frac{1}{2}l(l+1)|\xi_{\rm h}(r)|^2 \tag{3.159}$$

We can then define a mode inertia in a way that's analogous to a moment of inertia in basic mechanics, so

$$I_{nl} = \int_0^R \rho\left[\xi_{\rm r}^2 + l(l+1)\xi_{\rm h}^2\right]r^2 dr \tag{3.160}$$

Frequently the mode inertia is normalized by dividing by its value for the total mass and radius of the star, or

$$I(R) = M\left[\xi_{\rm r}^2(R) + l(l+1)\xi_{\rm h}^2(R)\right] \tag{3.161}$$

Low-degree modes have higher mode inertias than do high-degree modes, as do low-frequency modes compared to higher-frequency ones, which should agree with intuition.

3.9 Excitation and Damping

We've assumed that oscillations are adiabatic, but how good an assumption is that? One rough way to check is to compare the energy stored in a shell of the star to the energy that passes through it over a pulsational cycle. Let's first consider the Sun as an example, and look at the entire convection zone. Here the energy stored is just

$$U_{\rm layer} = \frac{3}{2}kT \times \frac{M_{\rm CZ}}{\mu m_{\rm H}} \tag{3.162}$$

where we've assumed an ideal gas, so $U = 3kT/2$ per particle. The mass of the solar convection zone is roughly 2% of the solar mass.

Compare this to the energy passing through the shell over the course of a typical oscillation period, which we can take as $\Pi \sim 300$s for solar p-mode oscillations. Since there are no nuclear energy generation sources in the convection zone, this is just

$$U_{\rm puls} = L_r\Pi = L\Pi \tag{3.163}$$

where L is the stellar luminosity. The ratio of the two is

$$\eta = \frac{U_{\rm puls}}{U_{\rm layer}} = \frac{2L\Pi\mu m_{\rm H}}{3kTM_{\rm CZ}} \sim 10^{-10} \tag{3.164}$$

Clearly adiabaticity is a good approximation here! As we approach the photosphere, it becomes less good: at the $\tau = 1$ level, $\eta \sim 0.1$, so in the outer layers adiabaticity fails.

If oscillations are adiabatic, we need a driving mechanism to supply energy to keep them going. Broadly speaking, there are two different families of driving mechanisms. Stars in (and near) the instability strip are driven by the heat engine mechanism as outlined in Section 2.1 and are intrinsically unstable; in a mathematical sense, amplitudes of oscillations grow exponentially until they reach the nonlinear regime, producing the high-amplitude pulsators we associate with this region of the HR diagram. Solar-like oscillators, on the other hand, are intrinsically stable, but are excited by broadband convective noise. Let's look at this stable case first.

A familiar (if simplified) example of this is the damped driven harmonic oscillator, where the driving force is some function of frequency ω. In one dimension, we can write

$$\ddot{x} = \zeta\dot{x} + \omega_0^2 x = f(t) \tag{3.165}$$

where $f(t)$ is the driving force. We can solve this in the usual way via Fourier transforms to get

$$x(\omega) = \frac{F(\omega)}{\omega_0^2 - \omega^2 + i\omega\zeta} \tag{3.166}$$

We measure the power, which we can get by multiplying through by the complex conjugate to obtain

$$|x(\omega)|^2 = \frac{|F(\omega)|^2}{\left(\omega_0^2 - \omega^2\right)^2 + \omega^2\zeta^2} \tag{3.167}$$

For a given damping parameter ζ, this has a maximum value when $\omega \to \omega_0$. If $F(\omega)$ is a broadband function, with power at many wavelengths, as would be anticipated when the white noise of convection is the driving force, then the amplitude of oscillations will be large where $\omega \approx \omega_0$ and we have *resonance*, and small elsewhere. If the driving force is broadband and there are multiple resonance frequencies (eigenfrequencies) within the range of driving frequencies, then all will be excited to large amplitude, which is what we see in solar-like oscillators. How large? It depends on the value of the damping constant ζ. In practice, of course, damping is relatively large, and the resulting amplitudes are small compared to those found in stars lying in the linearly unstable regime.

In such stars the amplitude of any oscillation will tend to grow with time. In practice, the star will have driving regions, which act to increase the amplitude locally, and damping regions, which act to decrease it. In order to have a sustained oscillation under these circumstances, the net driving must exceed the net damping over the entire star.

We can write the energy change as a function of time, dq/dt, for a parcel of gas as

$$\rho\frac{dq}{dt} = \rho\varepsilon - \nabla \cdot \mathbf{F} = \rho\left(\varepsilon - \frac{1}{\rho}\nabla \cdot \mathbf{F}\right), \tag{3.168}$$

where ε is energy input per unit mass from (generally) nuclear processes and $\nabla \cdot \mathbf{F}$ is the flux of energy through the boundaries of the parcel. If the term in parentheses is greater than zero, then the energy content is increasing. For efficient driving, the kappa mechanism must be inserting energy when the driving layer is undergoing maximal compression.

We can use what we know about opacity to put this into a more intuitive context. Consider two layers in a star, at radii r_1 and r_2, where $r_2 > r_1$. We assume the two layers have densities and temperatures (ρ_1, T_1 and ρ_2, T_2), and correspondingly differing opacities $\kappa(\rho, T)$ as well. Since the radiative flux is inversely proportional to the opacity, we can write

$$\delta F = -F\frac{\delta\kappa}{\kappa} = -F\delta \log \kappa = -F\frac{\partial \log \kappa}{\partial \log T}\frac{\delta T}{T} \tag{3.169}$$

The flux difference ΔF is the difference between the flux entering layer 1 and the flux leaving layer 2, so

$$\begin{aligned}\Delta T &= (F_1 + \delta F_1) - (F_2 + \delta F_2) \\ &= -F\frac{\delta T}{T}\left[\left(\frac{\partial \log \kappa}{\partial \log T}\right)_{r_1} - \left(\frac{\partial \log \kappa}{\partial \log T}\right)_{r_2}\right] \\ &= hF\frac{\delta T}{T}\left(\frac{d\kappa_T}{dr}\right)\end{aligned} \tag{3.170}$$

where in the last step we've defined

$$\kappa_T = \left(\frac{\partial \log \kappa}{\partial \log T}\right) \tag{3.171}$$

The condition for local driving then becomes

$$\frac{d\kappa_T}{dr} > 0 \tag{3.172}$$

which is generally fulfilled in the outer partial ionization zone(s) of classical pulsators.

However, for efficient coupling of the driving region to the star, we also want the thermal timescale of the driving region to be roughly the same as that of the oscillation periods, or

$$\Pi \sim \frac{3kTM_r}{2\mu m_{\mathrm{H}} L_*} \tag{3.173}$$

This increases with depth, implying that longer-period modes are driven by deeper layers than the short-period ones.

The transition region between adiabatic and nonadiabatic for an oscillation mode is set by where $\eta \sim 1$. Since η increases inwards, this implies that deeper modes are adiabatic and shallower ones more nonadiabatic. This matters, because if the transition depth for a mode lies *above* the partial ionization zone which could potentially drive it, then the mode is adiabatic in that zone, so effective driving doesn't occur. Conversely, if the transition depth like *below* the partial ionization zone, then the luminosity is effectively frozen in: energy leaks out too quickly for driving to occur. Modes for which the transition region is cospatial with a partial ionization zone are best-placed to be driven to large amplitudes.

We conclude with a couple of points to keep in mind. First, our simplified derivation of driving due to the kappa mechanism really provides only a necessary, and not a sufficient, condition for oscillations to be excited to visibility, because we haven't taken damping into account in detail. A highly-damped mode will not be visible no matter how efficient the excitation mechanism is: drawing on the one-dimensional damped driven oscillator analogy, we would be in the overdamped region of the solution.

Second, we have implied that in theory the kappa mechanism can't operate in convective zones, since small changes to the opacity in those regions have no effect because energy isn't primarily being transported there by radiation in the first place! The real picture can be somewhat more complex, because (1) some small portion of the energy *is* still being transported by radiation, and (2) changes in opacity and/or ionization state can change the convective character of the zone.

References

Aerts, C. 2021, RvMP, 93, 015001

Aerts, C., Christensen-Dalsgaard, J., & Kurtz, D. W. 2010, Asteroseismology (Dordrecht: Springer)

Appourchaux, T., Belkacem, K., Broomhall, A. M., et al. 2010, A&AR, 18, 197

Basu, S. 2016, LRSP, 13, 2

Basu, S., & Chaplin, W. J. 2018, Asteroseismic Data Analysis. Foundations and Techniques (Princeton, NJ: Princeton Univ. Press)

Bowman, D. M. 2020, FrASS, 7, 70

Chaplin, W. J., & Miglio, A. 2013, ARA&A, 51, 353

Christensen-Dalsgaard, J., Silva Aguirre, V., Cassisi, S., et al. 2020, A&A, 635, A165

Deubner, F.-L., & Gough, D. 1984, ARA&A, 22, 593

Garcia, R. A., & Stello, D. 2018, arXiv:1801.08377

Gough, D. O. 1993, 47th Session de l'Ecole d'Eté de Physique Théorique: Astrophysical Fluid Dynamics - Les Houches 1987, ed. J. P. Zahn, & J. Zinn-Justin (Amsterdam: Elsevier) 399

Kurtz, D. W. 2022, ARA&A, 60, 31

Pallé, P., & Esteban, C. 2014, Asteroseismology, Canary Islands Winter School of Astrophysics (Cambridge: Cambridge Univ. Press)

AAS | IOP Astronomy

Asteroseismology for the Nonspecialist

Derek L Buzasi

Chapter 4

Data Reduction Tools and Techniques

Stellar oscillations are accessible to our direct observation through their effects at the photosphere of the star, as the gas moves (in general, both radially and horizontally) in response to the acoustic waves. The two primary tools for the actual detection of stellar oscillations are those workhorses of astronomy, photometry and spectroscopy. Photometrically, the waves can change both R and T, but the latter is usually the dominant impact unless oscillation amplitudes become large, as with some classical pulsators such as Cepheids. In the case of purely radial oscillations, a general sense of how the oscillations map into velocity variations detectable by spectroscopy is intuitive, but for more complex oscillation modes we need to have recourse to a more formal mathematical description.

It's helpful to start off by discussing the approximate scale of the effects we are trying to detect, in terms of photometric amplitude, radial velocity amplitude, and frequency, and their dependences on basic stellar properties. Following Bedding et al. (2011), we treat the star as a blackbody and first consider the bolometric amplitude. For most oscillating stars, changes in radius associated with oscillations are very small, and for blackbodies the sensitivity of photometric amplitude is twice as large for changes in temperature as for changes in radius, so for simplicity we consider only the former. In that case, we have for luminosity,

$$L \sim R^2 T_e^4 \rightarrow \delta L \sim T_e^3 \delta T_e \tag{4.1}$$

or

$$\frac{\delta L}{L} \sim \frac{\delta T}{T} \tag{4.2}$$

If the oscillations are adiabatic, then we also have

$$P \sim \rho^{\gamma} \tag{4.3}$$

doi:10.1088/2514-3433/ae03a0ch4

so using an ideal gas equation of state where $P \sim \rho T$, we obtain

$$T \sim \rho^{\gamma-1} \tag{4.4}$$

We can then write

$$\frac{\delta T_e}{T} \sim (\gamma - 1)\frac{\delta\rho}{\rho} \tag{4.5}$$

For subsonic waves traveling in an adiabatic medium, we have

$$\frac{\delta\rho}{\rho} = \frac{\mathrm{v}}{c} \tag{4.6}$$

where c is the sound speed and v the wave velocity. The sound speed itself is related to the temperature (again assuming an ideal gas equation of state) through

$$c^2 = \gamma\frac{P}{\rho} = \gamma\frac{\rho k T}{\mu m_{\rm H}\rho} \sim \frac{T}{\mu} \tag{4.7}$$

We see the effects of the wave near the photosphere, where $\tau \sim 1$, so we take $T = T_e$. Putting Equations (4.2), (4.6), and (4.20) together, and neglecting changes in μ and γ, we then get

$$\frac{\delta T_e}{T_e} = \frac{\delta F}{F} \sim \frac{\mathrm{v}}{\sqrt{T_e}} \tag{4.8}$$

Real photometers have a bandpass, and thus are wavelength-sensitive, so we should consider the effects of observing at different wavelengths λ. If the star is a blackbody, then for small relative amplitudes, we have

$$\frac{\delta F_\lambda}{F_\lambda} = \frac{\delta B_\lambda(T)}{B_\lambda(T)} = \delta \log B_\lambda(T) \tag{4.9}$$

Taking the logarithm of the blackbody expression $B_\lambda(T)$ gives

$$\log B = K - 5\log\lambda - \frac{hc}{\lambda k T} \tag{4.10}$$

where K is a constant, so

$$\delta \log B_\lambda(T) = \left(\frac{hc}{\lambda k}\right)\left(\frac{\delta T}{T^2}\right) \tag{4.11}$$

This implies that

$$\frac{\delta F_\lambda}{F_\lambda} \sim \frac{1}{\lambda} \tag{4.12}$$

Thus, *ceteris paribus*, amplitudes are higher in the blue than in the red. Usefully for observing purposes, we can rewrite this expression as

$$\delta F_\lambda \cdot \lambda_{\rm bol} = \delta F \cdot \lambda \tag{4.13}$$

where λ_{bol} is the (temperature-dependent!) wavelength at which the *observed* amplitude is exactly equal to the *bolometric* amplitude:[1]

$$\lambda_{\text{bol}} = \frac{623 \text{ nm}}{T_e/5777 \text{ K}} \tag{4.14}$$

Combining this with our earlier expression (4.15), we can write an expression for the amplitude at any given wavelength:

$$\frac{\delta F_\lambda}{F_\lambda} \sim \frac{v}{\lambda T_e^{1.5}} \tag{4.15}$$

Observationally, a value of 1.5 for the exponent in the denominator works well for solar-like oscillations, while classical pulsators are better fit with a value of 2.0.

Calculating expected amplitudes is complex because it involves both excitation and damping processes, neither of which is known as well as we would like. Bedding et al. (2011) suggested, based on models by Christensen-Dalsgaard & Frandsen (1983), that

$$v \sim \frac{L}{M} \tag{4.16}$$

Observationally, a power law with an exponent s other than unity is likely to be a better fit, so we can generalize to

$$v \sim \left(\frac{L}{M}\right)^s \tag{4.17}$$

where authors have argued for a range of s typically in [0.7, 1.3]. In terms of near-surface quantities which are measurable using spectroscopy, we can write this relationship as

$$v \sim \left(\frac{T^4}{g}\right)^s \tag{4.18}$$

Converting the proportionality to an equivalency involves some uncertainty. We've already seen that the measured flux variability depends on the wavelength one measures at, and the same is true for the velocity. In practice, we measure the velocity using spectroscopy (see Section 4.4), which involves measuring velocity shifts caused by a potentially large number of spectral lines. Each of these lines has its own formation depth, and the velocity of the propagating oscillation is a function of depth. Helioseismic experiments take advantage of the high flux levels from the Sun to focus on one or a few specific lines, but asteroseismology experiments lack that advantage and therefore average over many. Typically, helioseismic experiments find the velocity of the strongest modes to be $\sim$20 cm s^{-1}, while because of this averaging asteroseismic instruments when used on the Sun typically measure

[1] See Kjeldsen & Bedding (1995) for a discussion of the genesis of these values

amplitudes ~10% lower. At a similar level of precision, the bolometric variability of the Sun due to the highest-amplitude oscillations is ~4 ppm. Combining these two gives a conversion factor of roughly 50 $\mathrm{ms^{-1}\,mmag^{-1}}$. Within a factor of ~2, this relationship is surprisingly good for most stars, though it fails when applied to high-amplitude classical pulsators.

4.1 Oscillation Frequencies

We've already seen the large frequency separation,

$$\Delta\nu = \left(2\int_0^R \frac{dr}{c}\right)^{-1} \tag{4.19}$$

which is the inverse of the sound travel time through the center of the star. For the Sun, observationally we find that $\Delta\nu_\odot = 134.9\ \mu Hz$.[2] Since the sound speed in an adiabatic medium is

$$c^2 \sim T \tag{4.20}$$

From Equations (1.4) and (1.5), we can determine the dependence of temperature on bulk values,

$$T \sim \frac{P}{\rho} \sim \frac{M^2/R^4}{M/R^3} \sim \frac{M}{R} \tag{4.21}$$

allowing us to write that

$$\Delta\nu \sim \left(\frac{M}{R^3}\right)^{1/2} = \rho^{1/2} = 135\ (\rho/\rho_\odot)^{1/2}\mu\mathrm{Hz} \tag{4.22}$$

so the large frequency separation (to first order) is proportional to the square root of the mean density of the star.

Typically multiple frequencies are excited simultaneously in a star. But where does the peak power lie? Which frequencies are most likely to be excited to large amplitudes? In the Sun the very name "five minute oscillations" implies that the frequency of peak power ν_{max} must be around 3 mHz; a more exact measurement gives 3.05 mHz.[3] How do we expect this to scale to other stars?

We can construct the expected scaling relation by considering a characteristic length and speed in the outer layers where we detect the oscillations. The characteristic length is just the pressure scale height

$$H = \frac{kT}{\mu m_{\mathrm{H}} g} \sim \frac{T}{g} \tag{4.23}$$

while the characteristic speed is just the sound speed. Dimensionally, from these we can construct a frequency

[2] Though see 7.1 for a more nuanced discussion.

[3] Though here again, see Section 7.1 for a more critical discussion.

$$\nu_{ac} = \frac{c}{H} \sim \frac{gc}{T} = \frac{g}{\sqrt{T}} \tag{4.24}$$

where we've used Equation (4.20) to make the last step. This is the acoustic cutoff frequency, which we met previously in Section 3.4.2. We now assume that the frequency of peak power ν_{max} scales with this characteristic frequency, so using the solar value to scale, we then have

$$\nu_{max} = 3.05\frac{g/g_{\odot}}{\sqrt{T/T_{\odot}}} = 3.05\frac{M/M_{\odot}}{(R/R_{\odot})^2\sqrt{(T/5777\ \mathrm{K})}}\ \mathrm{m}Hz \tag{4.25}$$

Along the main sequence this corresponds to ~0.36 mHz for an O star (corresponding to a period close to an hour) rising to >10 mHz for a late M dwarf (periods less than 2 min). Along the giant branch, as the radius grows, ν_{max} can fall as low as only a few μHz, corresponding to oscillation periods of days.

4.2 Spatial Response

To this point, we've treated the star as if it were a uniformly illuminated, radially pulsating sphere. Neither of these is of course the case! The disk is limb-darkened (or potentially brightened, depending on the wavelength of observation), and the surface inhomogeneities caused by oscillations are described by the spherical harmonics, $Y_l^m(\theta, \phi)$, which are in general non-radial. How large are these effects?

At some given time t, the intensity of the light measured from the star can be represented by an integration over the entire visible disk of the local intensity $I(\theta, \phi)$ at each colatitude θ and longitude ϕ

$$I = \int_A I(\theta, \phi)dA \tag{4.26}$$

For the moment, we neglect effects such as limb darkening and gravity darkening and weight all portions of the disk equally. As the star oscillates, the intensities of each dA vary according to the effects of the sums of all the oscillation modes, and become functions of time, as does the integrated light, so

$$I(t) = \int_A I(\theta, \phi, t)dA \tag{4.27}$$

We can describe the intensity on the surface of a spherical star using spherical harmonics, so that

$$I(\theta, \phi, t) = I_0(-1)^m c_{lm} P_l^m(\cos\theta)\cos(m\phi - \omega_0 t + \delta_0) \tag{4.28}$$

where the c_{lm} are the normalization constants discussed above, and take the values

$$c_{lm} = \frac{(2l+1)(l-m)!}{4\pi(l+m)!} \tag{4.29}$$

ensuring that the integral over the entire surface of each $|Y_l^m|^2$ is unity.

The observed intensity will then be the average of $I(\theta, \phi, t)$ over the projected area of the visible stellar disk, A

$$\bar{I}(t) = \frac{1}{A}\int_A I(\theta, \phi, t)dA \tag{4.30}$$

We can then integrate Equation (4.31), choosing for simplicity a spherical coordinate system whose polar axis points toward the observer. In this case the integral vanishes by symmetry for any $m \neq 0$, while for $m = 0$ it becomes

$$\bar{I}(t) = I_0 \cos(\omega_0 t + \delta_0) \cdot \left[2\sqrt{2l+1}\int_0^{\pi/2} P_l(\cos\theta)\cos\theta \sin\theta d\theta\right] \tag{4.31}$$

If we choose now to include limb darkening and/or gravity darkening, we simply add a limb darkening function inside the integral. The entire expression inside the square brackets is the *visibility function*, or the *spatial response* S_l^I. Note that if we are looking at oscillation *power* rather than amplitude, as is frequently the case, the relevant parameter is S_l^2 rather than simply S_l.

If we are interested in velocity rather than intensity, the process is the same, but now there's an extra factor of $\cos\theta$ in the integral to account for the fact that we are only sensitive to the portion of the velocity that is along the line of sight, so here

$$S_l^V = 2\sqrt{2l+1}\int_0^{\pi/2} P_l(\cos\theta)\cos^2\theta \sin\theta d\theta \tag{4.32}$$

where the superscript V indicates the spatial response for velocity.

Figure 4.1 illustrates the values of both S^I and S^V for a range of l. Including the effects of limb darkening affects the picture only modestly. As expected, averaging effects mean that stellar observations are only sensitive to oscillations of low degree l, typically $l \leqslant 3$, though spectroscopic (velocity) observations have some potential to be sensitive to modestly higher l modes.

The derivation above is for a particular alignment of the star, such that the polar axis is directed at the observer. To transform to other inclinations, we need to transform the coordinate system (Gizon & Solanki, 2003), such that

$$S'_{lm} = \Gamma_{lm} S_l \tag{4.33}$$

where the Γ_{lm} represents the rotation matrix. Details are given in Basu and Chaplin (2018); Christensen-Dalsgaard & Gough (1982).

We can of course combine the instrumental wavelength response and the visibility curve into a single scaling factor C_{nl} for flux or velocity amplitude, so that we can predict the anticipated flux or velocity amplitude for a given mode nl.

As implied above, the two main techniques for actually measuring stellar oscillations involve measuring periodic variability in either luminosity (flux) or radial velocity. The tools for these two approaches are photometry and spectroscopy, and we now move to discussing both approaches.

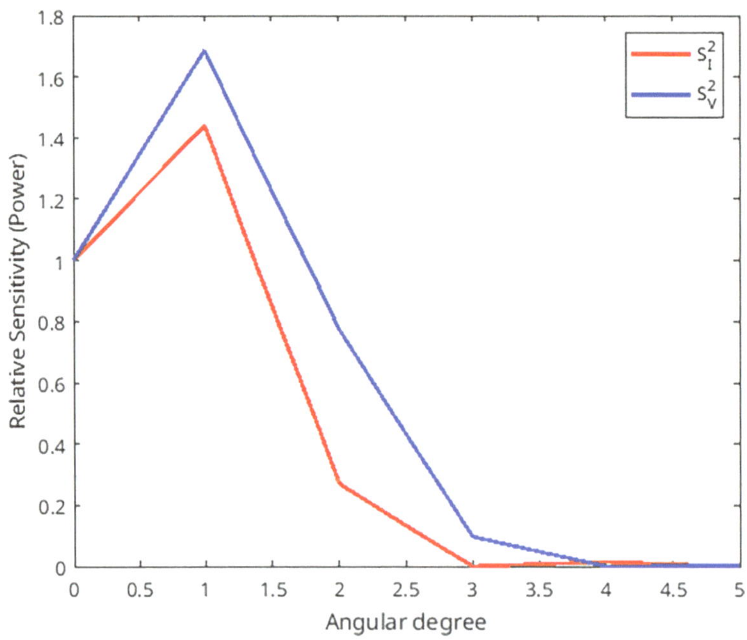

Figure 4.1. Relative sensitivities (spatial response functions) for photometric and velocity observations as a function of angular degree *l*, in the absence of limb or gravity darkening.

4.3 Photometry

"Photometry" describes the process by which we measure the light output of a celestial body such as a star. It can be done using a filter to limit the wavelength range of the light we measure, or in a broad-band way without a filter, though one should always remember that in the latter case we really still have a filter, only the filter bandpass is now given by the transmission and reflection characteristics of the optical train along with the wavelength response of the detector. Asteroseismically-driven photometric variability is generally at the level of 1% or less—**much** less for solar-like oscillators—so typically such photometric measurements are performed using a space platform rather than a ground-based telescope, though part-per-thousand (ppt) photometry is becoming much more routine from the ground, particularly with the advent of diffusers (Stefansson et al. 2018a, 2018b).

Here we give only a high-level view of the process of photometry, since many detailed references exist (see, e.g., Jenkins et al. (2010), Vanderburg & Johnson (2014), Lund et al. (2015), Aigrain et al. (2016), Huang et al. (2015), Van Cleve et al. (2016), Libralato et al. (2016), Pope et al. (2016), Buzasi et al. (2016), and Petigura et al. (2018) for Kepler/K2, and Lightkurve Collaboration et al. (2018), Oelkers & Stassun (2018), Feinstein et al. (2019a, 2019b), Nardiello et al. (2019), Woods et al. (2021), Plachy et al. (2021), Huang et al. (2020a), and Handberg et al. (2021) for TESS), and focus on those details specific to space-based asteroseismic photometry, particularly that deriving from the Kepler and TESS missions. The most important parts of the process from the perspective of a user of the data are background estimation and subtraction, extraction of the scientific signal, and correction of instrumental effects. We will illustrate the process below using several different photometric pipelines, but will primarily focus on TASOC (Handberg et al. 2021; Lund et al. 2021) and *eleanor* (Feinstein et al. 2019a, 2019b). Custom light curves can also be created using the *lightkurve* tool (Lightkurve Collaboration et al. 2018).

Sky background flux reflects a combination of inputs including scattered sunlight, moonlight, and earthlight along with zodiacal light and contributions from unresolved background sources. The simplest approach is just to take the faintest nearby pixels and assign them to represent the background level for each frame; this can work surprisingly well but finding the correct balance between too few pixels (with a correspondingly low SNR for the background estimation) and too many (risking contamination by background stars) is difficult, and a poor choice of background pixels can also risk mischaracterizing complex sky background structure in the frame. Traditional ground-based photometry performs background characterization through the use of a pixel annulus surrounding the target, placed far enough from the target pixels to avoid most contamination. Rather than calculating an individual annulus for each source in this way, the pipeline for the Kepler mission constructed dedicated small (2×2 pixel) background frames scattered over each CCD, and interpolates a 2D polynomial fit to the background derived from these to apply a background correction to the entire field. The locations of these background estimator frames are chosen to avoid bright stars and bad pixels, and are concentrated toward the edges of CCD fields to better characterize edge effects. These background estimates were performed at long-cadence, and were further interpolated to short cadence time steps in order to apply them to the short cadence data. TESS adopted an approach more similar to that used for ground-based photometry, estimating a background level for each observation using a group of background pixels peripheral to the target pixels, with some judicious editing of the annulus to avoid contamination by other stars in the field; the latter approach is not always successful. Figure 4.2 shows the typical result for a star in a relatively uncrowded field.

As intimated above, proper background characterization is typically complicated by the fact that the large field of view of the various space missions leads to varying background levels across the field of view, so a true background characterization requires two-dimensional (plus time!) fitting of background levels. Figure 4.3 shows a snapshot in time of the background level for a single TESS detector, and the results of the TASOC background characterization and removal process. In this case, the overall background has been fit by a relatively complex algorithm involving the masking out of bright stars followed by sigma-clipping and an interpolation of a sliding fit of the remaining flux. However, in the case of TESS even this procedure does not account for the so-called "cornerglow" present in corners of each CCD farthest from the center of the field (Handberg et al. 2021), which requires its own specialized approximation which is not entirely successful in practice at removing the effect.

In addition to "cornerglow" and other complex structures in the background, other complications in background estimation include:

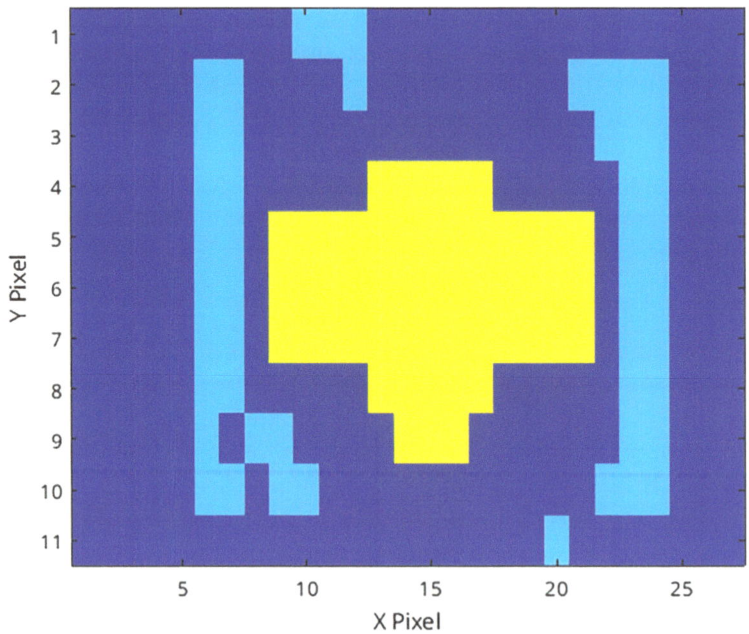

Figure 4.2. A typical TESS "postage stamp" frame, measuring 11 × 27 pixels in size. The yellow pixels are assigned to the target star (31 Aql), while the light blue pixels are used to evaluate the background. The irregular shape of the background is intended to avoid contamination by brighter background stars or other sources.

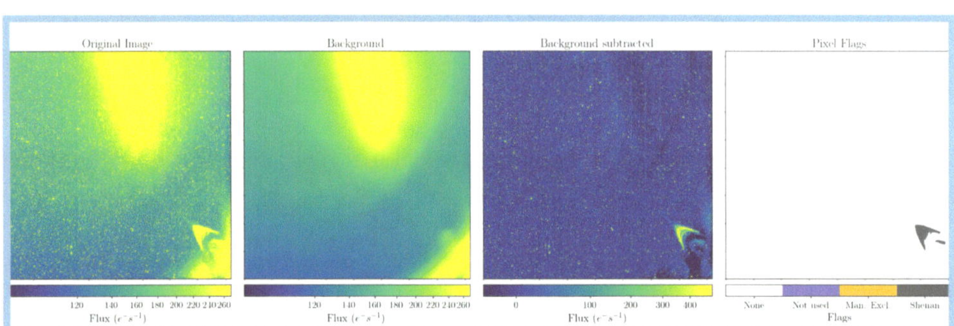

Figure 4.3. An example of background extraction in FFIs, here for Sector 1, Camera 1, CCD 2. Panel 1 (leftmost): the raw calibrated FFI image. Panel 2: the estimated background including both the broad background and the corner glow. Panel 3: the background-corrected image. Panel 4 (rightmost): identification and flagging of the residual "cornerglow" feature in the background-corrected image. Reprinted with permission from Handberg et al. (2021).

- Cosmic rays: Most spacecraft are affected by the flux of high-energy particles arising from the Sun and the Galaxy. Cosmic rays are sufficiently energetic to fill the wells of at least several CCD pixels, and can arise from any spatial direction. Satellite pipelines include cosmic ray identification and rejection algorithms, and will flag affected frames, but these algorithms are imperfect.

- Bright objects: Observing near bright objects such as Venus, Mars, the bright Earth, and the Moon can lead to variable and complex background structures, particularly if the light from these objects is scattered and reflected by the optical train.
- Excess flux: Near bright stars, some of the signal within the aperture arises from the wings of the images of nearby sources, contaminating the signal from the target. Bright variable sources in or near the field of view can further complicate background estimation. In addition, for extremely bright sources, the majority of the pixels in the postage stamp, or even all of them, may be contaminated by light from the broad wings of the primary source, making precise background estimation impossible.
- Argabrightening events: These are occasional diffuse illuminations of the focal plane, which last a few minutes, and are of unknown origin. Typically these events are flagged in the light curve.
- Loss of guidance: At times, the spacecraft can lose fine guidance, or more rarely lose guidance altogether. The latter generally is flagged and should be discarded, while the former data can sometimes be used with care, but if the increased jitter is large enough stellar light can contaminate background pixels.

TESS full-frame images (FFIs) are contaminated by light from background sources (e.g., zodiacal light, scattered light from solar system objects). This is of particular concern for TESS as it lies on an orbit that takes it close to earth periodically, causing large-scale changes in background flux.

Handberg et al. (2021) tested, assessed, and compared several background estimation algorithms for TESS data. In each case, the algorithms were applied to the entire CCD frame rather than simply a single postage stamp. Approaches included:

- Row-by-column estimate: Here the background was estimated individually along each row of pixels and each column. Each column or row was then divided into 60 bins, smoothed with a Hamming filter based on the 10th percentile value within each bin, and then the full 2-D background estimated by averaging the row and column estimates. This performed well at the center of the frame, but tended to undercorrect at the edges.
- Binned Mode Estimate: The frames were divided into 128×128 pixel subimages, and the mode for each subimage estimated using a robust algorithm. The resulting modal image was then fit with a two-dimensional polynomial to estimate the background. Overall, this approach did not work well, particularly on the cornerglow.
- Kepler-style: Inspired by the approach used for the Kepler spacecraft, in this implementation Handberg et al. (2021) tiled the CCD with 9×9 pixel subimages, calculating the mode of each subimage, and then interpolating across the complete tiling to estimate the background. Sigma-clipping of the mode array was used to make the background smoother, and in an attempt to mitigate the problems with the previous approach, the density of points for

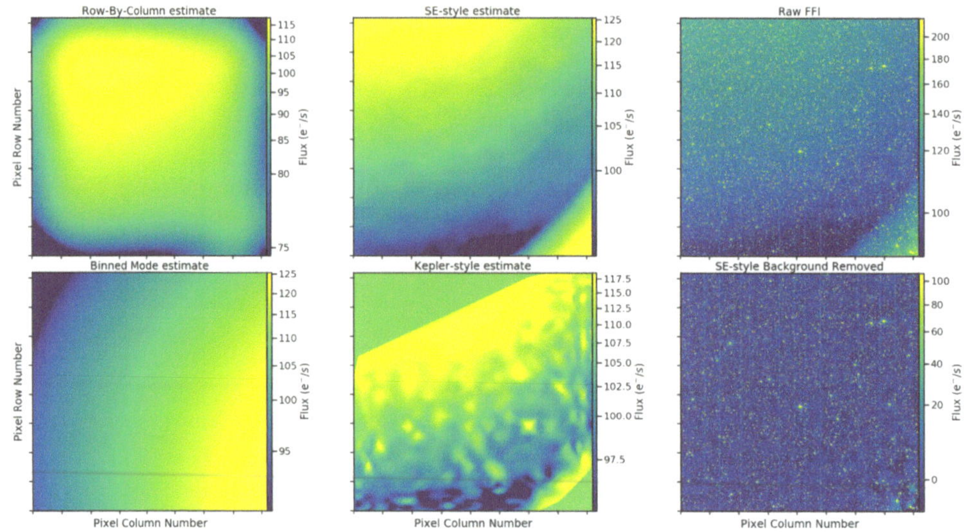

Figure 4.4. Background estimates of an FFI from TESS Camera 1, CCD 2 in Sector 3, for each of the four estimation methods described in the text. The raw and background-subtracted images are shown on the right. In this case, the SE-style estimate removes the background most robustly. Reprinted with permission from Handberg et al. (2021).

the interpolation was increased near the edges of the field. As with the previous approach, this tends to fail near the edges of the field, and in areas where sharper features are present.

- The last approach implemented a version of the SourceExtractor (Bertin & Arnouts 1996) algorithm. The full image was divided into 64×64 pixel subimages, after 3-σ clipping to mask stellar images. The mode of each image was estimated, while the cornerglow was treated separately through a second subdivision using radial rather than Cartesian coordinates, and the two approaches performed iteratively. The resulting modal array was then processed using a 3×3 moving median filter, and reinterpolated back to the original image size.

Handberg et al. (2021) found that the cornerglow caused difficulties with algorithms relying on smoothing or interpolation, due primarily to its sharp edges. Overall, the last approach was fastest and most effective both overall and in treating the cornerglow. Figure 4.4 shows a comparison of the different approaches.

Following background characterization and subtraction, an initial light curve is typically created by combining the pixels containing light from the star. The most common approaches here include:

- Mirroring the typical ground-based approach by summing the pixels contained in a circular aperture (or at least approximately circular, given the small number of pixels typically included in each). This has the merits of speed and simplicity, but determining the size of the aperture can add a complicating factor. The *eleanor* algorithm simply uses a fixed set of apertures and chooses the one which produces the "best" light curve by their metrics (Figure 4.6; see Feinstein et al. 2019a, 2019b). The optimal aperture for a target can depend on the stellar magnitude, the background and noise levels, the presence of nearby bright stars ("crowding"), and variations in the pixel response function as a function of both location in the field and time. In addition the centroid location for the star can vary periodically due to the spacecraft orbit as well as pointing instabilities. Generally speaking, brighter stars require larger apertures (Figure 4.5). Furthermore, the relatively small numbers of pixels in the optimal aperture typically leads to aperture shapes that depart significantly from symmetry, particularly for fainter stars, and the desire to avoid inclusion of nearby stars in observations of clusters or crowded galactic fields can lead to decidedly suboptimal aperture sizes.

- An improved option can involve the use of asymmetrical apertures. Thus, the K2P2 pipeline developed for K2 (and later extended for use with TESS) combines a clustering algorithm guided by the known locations of targets in each postage stamp combined with a "watershed" algorithm which makes use of (inverted!) basins surrounding each star to derive an optimal set of pixels to use for photometric extraction (Lund et al. 2015). This approach does not specify in advance either the shape or size of the aperture, so in theory is both more flexible in general and superior in crowded fields, but is significantly more computationally intensive. An example of the process applied to TESS photometry is shown in Figure 4.7. Alternatively, an aperture can be constructed one pixel at a time, maximizing the SNR or some other figure of merit at each choice of additional pixel (Buzasi et al. 2016; Nielsen et al. 2020).
- Finally, the most sophisticated approach (but most costly) is to perform true PRF-based photometry, making use of the instrumental response function to

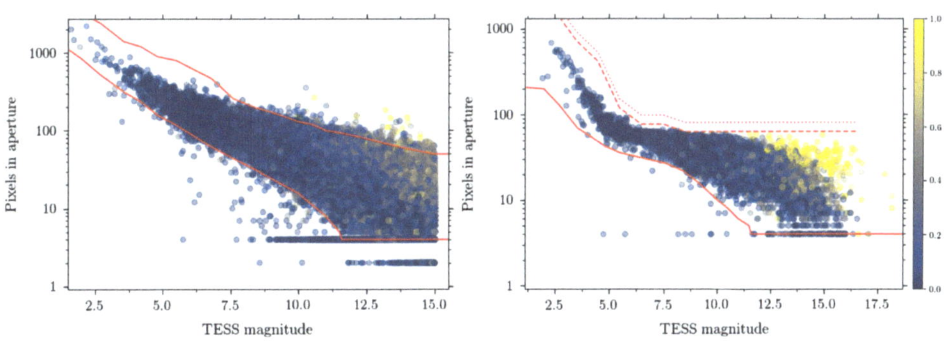

Figure 4.5. The optimal number of pixels in TESS apertures as a function of TESS magnitude, based on data from TESS Sector 1. The left panels show apertures for 1800 s (long cadence) targets, while the right panels show apertures for 120 s (short cadence) targets. Individual points are color-coded to indicate estimated contamination levels from nearby sources. The full red lines give boundaries for data validation, while the dotted line in the right panel gives the median relation for the maximum allowed number of pixels given the stamp size, and the dashed line shows the corresponding relation under the assumption that the aperture is circular. Reprinted with permission from Handberg et al. (2021).

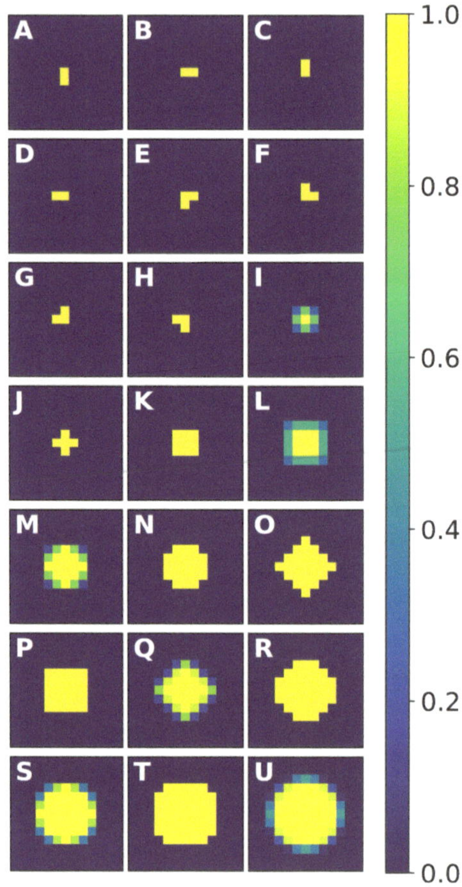

Figure 4.6. Default library of apertures *eleanor* selects from to create light curves. Within an aperture, pixels can be weighted between 0 and 1, as indicated by color. In crowded fields, only apertures with aperture pixel sums less than 9 (apertures A–J) are used to extract light curves. Reprinted with permission from Feinstein et al. (2019a).

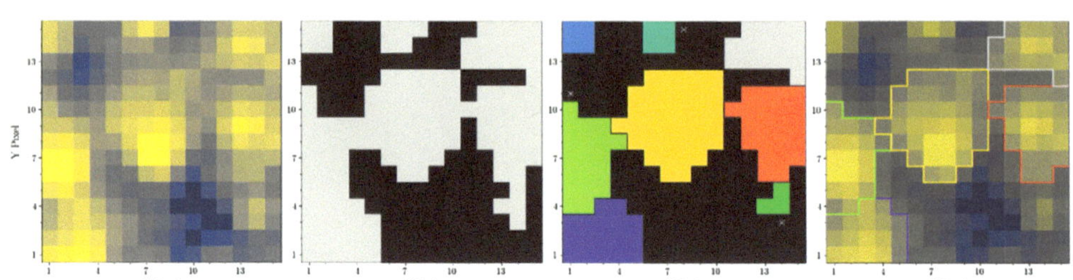

Figure 4.7. Example of aperture definition by the TASOC pipeline for a target in a relatively crowded region, here for the star TIC 410446575 in Sector 3 (central yellow aperture; TIC stands for the TESS Input Catalog (Stassun et al., 2018, 2019). From left to right, the first panel shows the summed image of the pixel cutout used for the star, the second panel gives in gray the pixels with flux levels above the adopted threshold, the third panel shows the apertures found from applying the clustering and watershed algorithms, and pixels identified as noise are marked with a white "×", and the last panel shows the outlines of the final aperture for the cutout after discarding apertures with 4 pixels or less. Reprinted with permission from Handberg et al. (2021).

> a true point source. In the case of Kepler, PRF data were collected prior to launch, but the PRF can vary significantly over time, so an improved model makes use of the more than 1.1 million sources contained in the Kepler FFIs; Martínez-Palomera et al. (2023) made use of these data to produce 1080 PRF models—one for every channel and quarter of interest—at the cost of some $\sim 10^5$ s of CPU time. These PRF models are then combined with *Gaia* source location information to extract source photometry.

Classical pulsators, with their large-amplitude variations, pose particular challenges to photometric extraction software. A particular aperture size and shape that may be appopriate to a Cepheid or RR Lyrae star at minimum light could well be oversized at maximum, and constructing an aperture based on the mean image of the star, as many pipelines do, will lead to a mismatch at maximum and minimum. The solution is either a varying aperture size, which can lead to its own problems, or to choose an aperture appropriate to maximum brightness, accepting that it will likely introduce modest additional noise at minimum (Plachy et al. 2019; Bódi et al. 2022).

Whatever process is used to obtain a raw light curve from the pixel data, it is almost certain that the result will include both the stellar signals which are of interest **and** signals deriving from the spacecraft and its environment. The latter include such contributions as spacecraft pointing jitter, residual scattered light, thermal drifts that lead to changes in pixel size or camera focus, and other periodic or quasi-periodic behaviors in either the camera or the spacecraft as a whole. Positional changes on the detector map to changes in photometry due to sub-pixel response structure in the detector.

The simplest approaches to characterizing and removing instrumental contributions to the measured light curve begin with determining the centroid of the stellar image and making use of a linear or low-order polynomial fit between this centroid and the various undesired contributions to the light curve in order to subtract or divide them out of the final signal.

A more complex approach makes use of co-trending basis vectors (CBVs), which are a set of orthonormal basis vectors that represent the shared variability among all or most stars in the field. The expectation is that such shared variability is likely to represent instrumental rather than stellar effects, as the stars do not "know" about one another. This common variability is then removed from the signal, leaving only the stellar signal behind. This process has complications, however. Fitting the entire instrumental variability signal can require a substantial number of basis vectors (see Figure 4.8), and overfitting can rapidly become a very real concern, so choosing the "correct" number of basis vectors to use can become more art than science. In addition, the fitting of each basis vector itself necessarily is an imperfect process which introduces noise, so the best number of basis vectors can also vary as a function of the SNR of the source itself. For example, *eleanor* (Feinstein et al. 2019a) uses the the three most significant CBVs as a default, though users can opt for more or fewer. Basis vectors are normally fit by scaling each and subtracting it from the flux time series using least-squares minimization, though more sophisticated algorithms can also be used. In practice, rules of thumb are often derived to guide

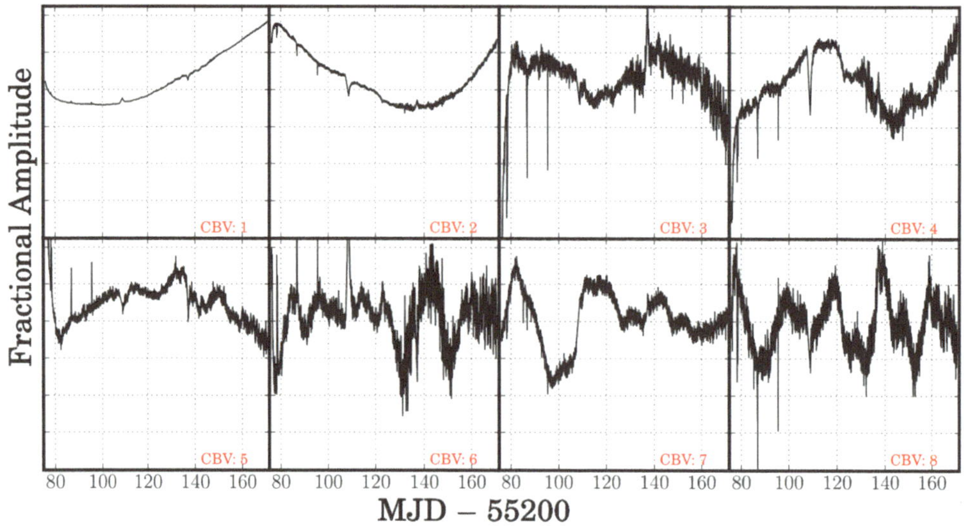

Figure 4.8. The first eight CBVs from Channel 50 in Quarter 5 of Kepler. These are the CBVs with the largest contribution to the systematic instrumental variability. From https://nexsci.caltech.edu/workshop/2012/keplergo/PyKEprimerCBVs.shtml.

the choice of CBVs to apply to large data sets, while correction of individual targets (using, e.g., *lightkurve;* Lightkurve Collaboration et al. 2018) is usually approached by starting with a small number of CBVs and increasing until, in the subjective judgment of the user, the fit achieves an optimal balance between systematics removal, overfitting, and the introduction of unwanted noise. It is frequently difficult or impossible to achieve a satisfactory result with classical pulsators and other targets with large amplitude variability.

An alternative way of removing instrumental systematics from the light curve derives from the "ensemble photometry" approach outlined by Gilliland & Brown (1988). In this method, the light from a number of nearby stars is combined to produce a kind of super-comparison star; if the number of stars chosen to constitute the comparison is large enough (and stars with large variability are avoided) then their individual stellar signals will cancel, and variations in the comparison light curve will be dominated by instrumental systematics (and weather and transparency variations, in the ground-based implementation). A difference between this and the CBV method described above is that the "ensemble" approach is more representative of local variations, and in practice can work better for highly variable sources. An example of this implemented for the TASOC light curves is given in Handberg et al. (2021) and shown in Figure 4.9.

An additional complication can arise from extremely bright stars. Such targets can locally saturate the detector, leading to flux bleeding into neighboring pixels and in particular down CCD columns. In practice, for space missions such as Kepler and TESS, the flux is conserved, so increasing the number of pixels contained in the aperture used can capture all of the flux. However, when the apertures continue to extend along the columns with the goal of capturing all of the photoelectrons, they

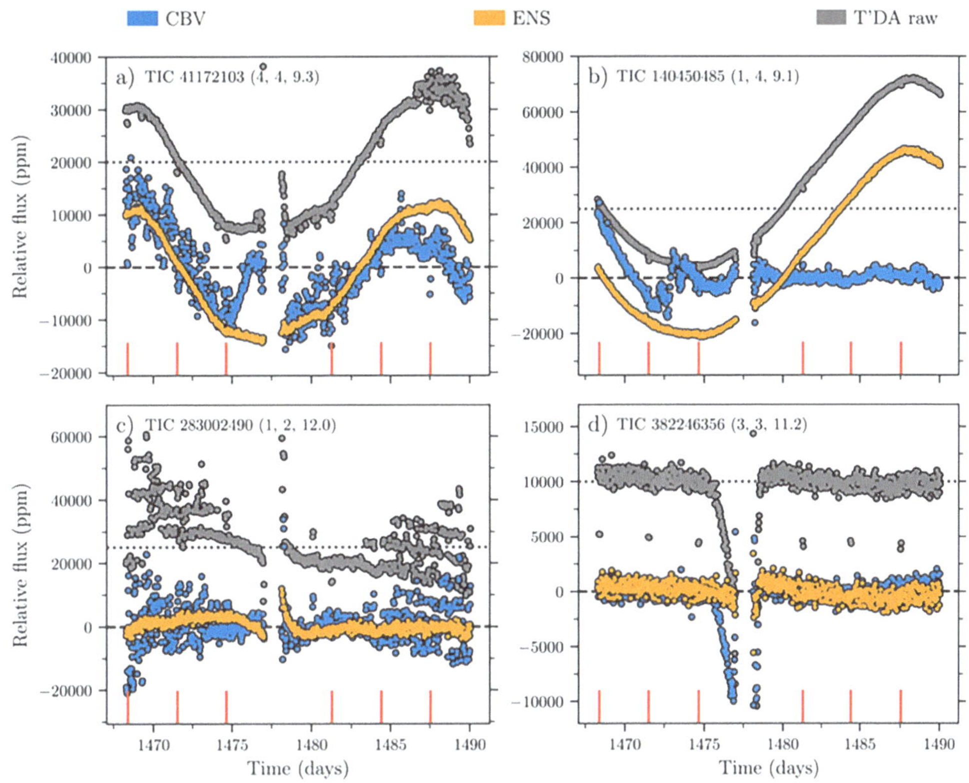

Figure 4.9. Example light curves for stars observed in 1800 s during Sector 6, which highlight the particular strengths of the ensemble (ENS) method. The TIC number of the targets is given at the top of the panels, including in parentheses the target camera, CCD, and TESS magnitude. The color of the points/lines indicates the source of the data. The raw light curves have been offset vertically from the corrected light curves by the difference between the horizontal dashed and dotted lines. The vertical red lines in the left panels mark the times when the spacecraft reaction wheel is desaturated, which has a strong negative impact on pointing precision. Reprinted with permission from Handberg et al. (2021).

can reach the edge of the postage stamp region or even the detector itself. In those cases, aperture photometry will fail to capture the entire flux, though sophisticated techniques such as "halo" photometry have been developed to reconstruct the missing flux based on the wings of the PSF (White et al. 2017 and Figure 4.11). A further complication for bright stars near the end of the detector itself arises from the "smear correction" routinely performed to correct the measured flux for the image smear caused by the fact that the camera lacks a shutter. Again, relatively sophisticated techniques can be used to compensate for this effect (Pope et al. 2016), but the user should be aware that additional effort is required in these cases.

High-amplitude pulsators pose particular problems, because the amplitude of the variability is generally large compared to the systematic effects of the spacecraft and instrument, making the latter hard to identify and remove. The impact of traditional detrending approaches varies, but can include the addition of amplitude changes (mimicking real physical effects like the Blazhko effect) or even removal of most or all of the pulsation signal. Accordingly, specialized pipelines are required in these cases (Bódi et al. 2022; Plachy et al. 2019, 2021). The baseline pipeline lightcurve products for both Kepler and TESS provide two different light curves, one based on simple aperture photometry (SAP), which represents a simple sum over pixel values in the chosen photometric aperture, and another labeled "Pre-search Data Conditioning" (PDC-MAP for Kepler and PDC-SAP for TESS; Stumpe et al. 2012), which attempts to fill gaps and correct offsets, and apply CBVs to remove instrumental effects. PDC light curves are specifically optimized for exoplanet searches, and have shown a tendency to remove longer-period behavior whether its origin is instrumental or stellar. Despite the nominal noise level in the PDC curves generally (though not always!) appearing better than the SAP ones for asteroseismology, it is best practice to check both light curves for each object. This is particularly true for stars with longer-period pulsations, such as massive stars and red giants. An illustration of different forms of light curve outputs from *eleanor* (Feinstein et al. 2019b) is shown in Figure 4.10.

Gilliland et al. (2011) examined on-orbit noise properties for the Kepler mission, decomposing noise for a sample of 12th magnitude stars into temporal, instrument, and intrinsic stellar terms. For such stars, the anticipated noise from count statistics and CCD read noise is approximately 14 ppm. However, additional instrumental noise sources added another 10 ppm, while the stars themselves, in the form of granulation and activity, added another 10 ppm, leading to overall noise levels

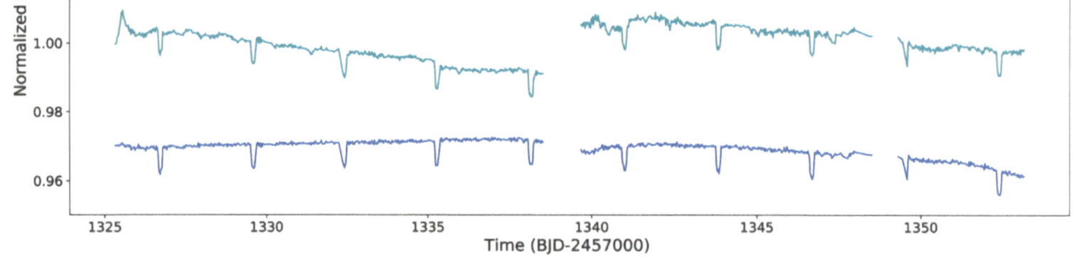

Figure 4.10. Example of the four types of light curves that can be extracted and modeled using *eleanor*. Raw flux is the sum of pixels in the aperture. Corrected flux is raw flux with a linear regression as a function of pixel location, background, and time. PCA flux is a Principal Component Analysis subtracted flux, to remove common systematics between targets on the same camera. PSF-modeled flux is the 2D Gaussian PSF-modeled flux. The 10 transits for WASP-100 are clearly seen in all four light curves. The visible gaps in the light curves result from times when the spacecraft is reoriented in order to downlink data. Reprinted with permission from Feinstein et al. (2019a)

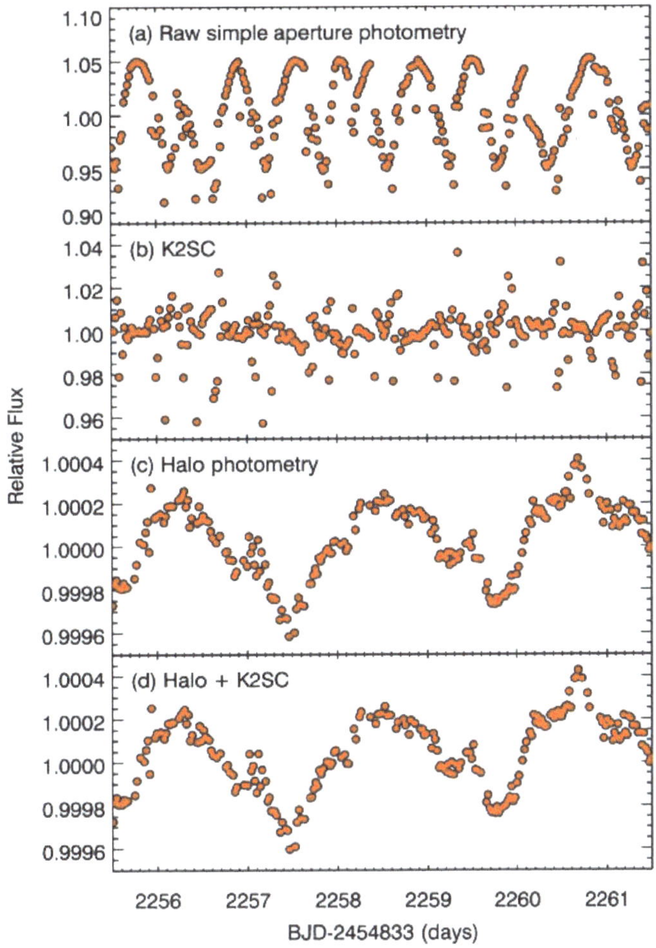

Figure 4.11. Six-day segment of the bright Pleiades star Alcyone light curve processed with different methods: (a) raw simple aperture photometry, (b) the K2SC pipeline (Aigrain et al. 2016), (c) halo photometry and (d) halo photometry with post-processing with K2SC. Note the change in scale between panels. Reprinted with permission from White et al. (2017).

(when sources are added in quadrature) of approximately 20 ppm, some 40-50% higher than initially anticipated. Results for TESS are broadly similar (Huang et al. 2020b and Figure 4.12), with achieved noise (using in this case the QLP pipeline; Huang et al. 2020c) only modestly above expectations based on theory (Sullivan et al. 2015). Overall, photometric noise levels in modern space-based instruments appear to closely approach theoretical limits, particularly for well-separated unsaturated targets at high frequencies, though instrumental effects remain meaningful in other cases.[4]

[4] Always *look* at your data, preferably after processing with more than one pipeline.

4.4 Spectroscopy

In addition to the brightness changes associated with stellar oscillations, we also expect the presence of velocity variations associated with the line-of-sight projection of the oscillatory motion. For stars like the Sun, the velocities associated with seismic oscillations are of the order of 10 cm s^{-1}. Pressure-mode oscillations give rise to surface motions, which are predominantly radial, so the contributions in integrated light are confined to regions near the center of the disk, which is turn implies less spatial averaging than is seen in photometric measurements; radial velocity data are therefore somewhat more sensitive to modes of higher angular degree l than are photometric data. A new generation of ground-based echelle spectrographs is available, offering precision at the < 1 m/s level, including ESPRESSO (Pepe et al. 2021), EXPRES (Petersburg et al. 2020), KPF (Gibson et al. 2016), MAROON-X (Seifahrt et al. 2018), and NEID (Halverson et al. 2016).

The most direct approach for the spectroscopic detection of oscillations, pioneered by Brown et al. (1991) is the measurement of integrated velocity using the Doppler effect, by means of cross-correlation of time-series spectra against one another. As is the case for photometric detections, the minimum detectable surface velocity is limited by photon noise, and in this case is given approximately by Peri (1995)

$$\delta V_{\rm rms} = \frac{K_{\rm C}}{Rn^{1/2}d}\frac{1}{{\rm SNR}N^{1/2}} \tag{4.34}$$

Here SNR is the signal-to-noise ratio, n represents the number of resolution elements across a typical spectral line, R is the resolving power of the spectrograph, d is the

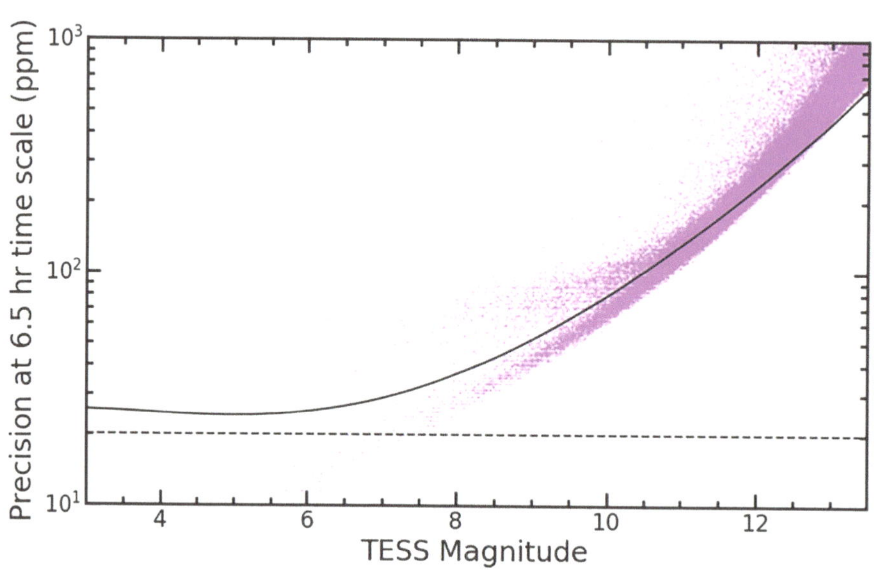

Figure 4.12. TESS noise precision based on QLP data for approximately 10 million targets from the first 2 years of observations. Noise is evaluated at a 6.5-hr timescale, and the solid line shows the predicted photometric precision assuming Gaussian noise. Reprinted with permission from Huang et al. (2020b).

typical relative depth of a spectral absorption line, and N is the number of accessible spectral lines. K_C is a shape parameter of order unity accounting for the fact that sharp absorption lines are easier to measure. Note that K_C decreases for lines which are Doppler-broadened, so slowly-rotating stars are more accessible.

Alternatively, one can look at changes in the line shape or centroid. We can intuitively understand the cause of these changes by realizing that, as optical depth varies across the line profile, we are sampling different depths in the stellar atmosphere. Thus, as motions caused by an oscillation propagate through the atmosphere, they map into changes in the line profile. Similarly, local periodic variations in other physical parameters, whether temperature, surface gravity, or abundances, will also map to changes in the line shape. One approach is illustrated in Figure 4.13; here the mean line profile for the δ Scu star HD 41641 has been subtracted from the time series spectra, and the residuals for each observation converted into a phased image, making the profile variations due to oscillations obvious (Escorza et al. 2016). Searching for changes in the equivalent width of temperature sensitive lines like Hα, as done by Bedding et al. (2001) for βHyi, is a related approach.

What is the scale of these variations? For classical pulsators, such as Cepheids or δ Scuti stars, we find velocity variations typically of tens of kilometers per second associated with the oscillation of a single mode or two (Figure 4.13). At the other extreme are solar-like oscillations, where while the maximum total velocity excursions can total up to a few hundred meters per second, the large number of modes typically excited implies that the velocity range for a single mode is a few tens of cm s^{-1} or less.

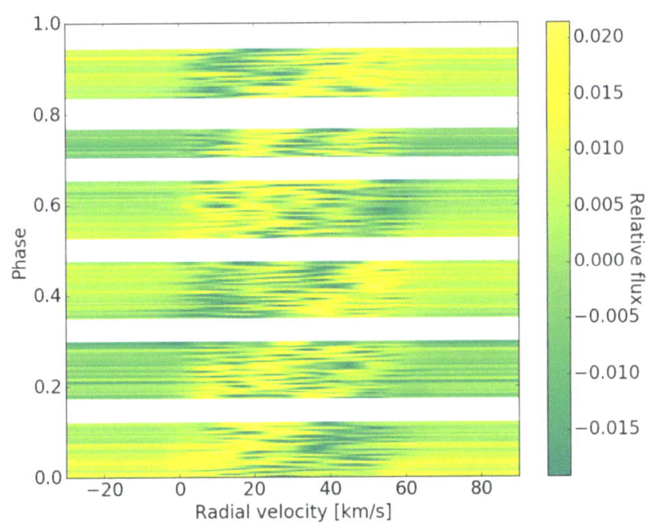

Figure 4.13. Intensity diagram of the residual flux after subtracting the mean line profile from each individual profile for a time series of the δ Scu star HD 41641. The result is then folded in phase with an oscillation period, making the periodic line profile variations readily apparent. Reprinted with permission from Escorza et al. (2016).

The instrumental requirements for these two cases are clearly substantially different. If we conservatively assume that we we have sufficient SNR to adequately determine the centroid of a single line to ~0.01 resolution elements, then to adequately sample the velocity range of a classical Cepheid using only that line might require a resolution of

$$R = 0.01\frac{c}{\Delta v} = \frac{3 \times 10^5\,\mathrm{km\,s^{-1}}}{30\,\mathrm{km\,s^{-1}}}\frac{\lambda}{\Delta\lambda} \sim 10^2 \tag{4.35}$$

whereas a similar calculation in the case of a solar-like star implies a required resolving power $\sim 10^5$ times greater, or $R \sim 10^7$, which is clearly impractical. Fortunately, however, we are not limited to using individual lines! In practice most spectrographs used for this purpose are multi-order echelle spectrographs with large spectral ranges, as much as the entire visible spectrum from 400–700 nm, or more. In such a case one can simultaneously make use of a large number of lines; the solar spectrum, for example, contains of order $\sim 10^5$ lines, implying that the required resolving power might be of order

$$R \sim \frac{10^7}{\sqrt{10^5}} \sim 3 \times 10^4, \tag{4.36}$$

which is indeed the order of magnitude of the resolving power of instruments that have successfully resolved solar-like oscillations. Of course, signal-to-noise ratio is also important, and the combined requirements of high SNR and large resolving power drive the user to large-aperture telescopes for solar-like stars, and even then the stars need to be quite bright (consider the example targets listed in Chapter 2). Since such stars are widely separated on the sky, and in any case the instruments in question are generally fed by a single optical fiber, radial-velocity asteroseismology is generally restricted to a single star at a time, making it considerably less efficient from an observing point of view than is space-based photometry, which can simultaneously observe large numbers of targets. Another issue is that ground-based observations from a single site can be performed only at night and in good weather, leading to a low duty cycle and a poor window function, though this can be obviated by adopting a multi-site approach, as was done in the past by the Whole Earth Telescope (WET; Provencal et al. 2014) for white dwarf observations and is currently being done for solar-like oscillation observations with the Stellar Oscillations Network Group (SONG) observatory network (Figure 4.14; Frandsen et al. 2018).

Given this limitation, what then in the modern era is the attraction of performing asteroseismology using high-precision spectroscopy? One advantage is that the contribution to granulation noise is smaller in radial velocity measurements than in luminosity. Figure 4.15 shows measured granulation noise for the Sun and two solar-like stars in both radial velocity and luminosity; the latter is an order of magnitude larger compared to the anticipated amplitudes of p-modes in these stars. A second advantage is that longer observing baselines are possible for well-chosen objects, and observations can be coordinated between multiple instruments at

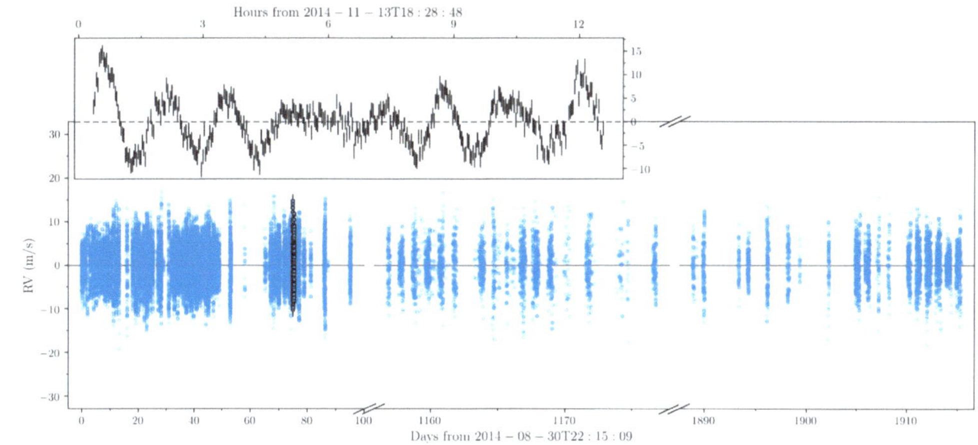

Figure 4.14. Radial velocity measurements from three different seismic SONG campaigns on γ Cep, presented after filtering and subtracting a nightly median. The error bars in the inset represent one night of observations from the 2014 campaign. The data from the inset are displayed in the main plot as blue circles with black edges and error bars. Reprinted with permission from Knudstrup et al. (2023)

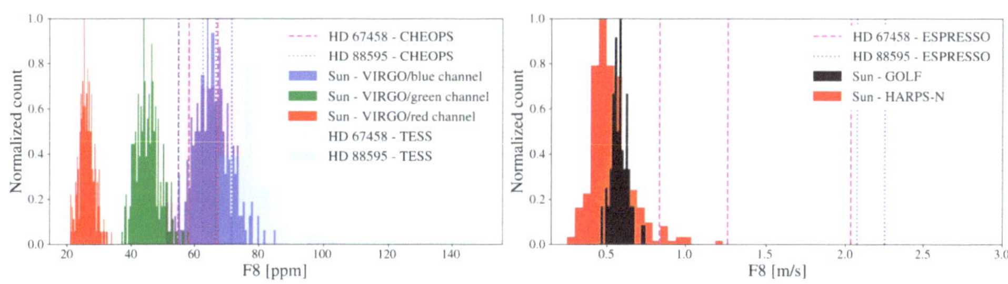

Figure 4.15. Left: Comparison of the stellar signal amplitude at short timescale for different stars (the Sun, HD 67458, and HD 88595), using a metric (F8) derived from a frequency region where the granulation signal is dominant. Shown are F8 metric distributions for VIRGO observations of the Sun (red, green, and blue histograms) and CHEOPS observations of HD 67458 and HD 88595 (colored dashed and dotted vertical lines, respectively). F8 values computed from TESS light curves are also shown (magenta and cyan histograms). Right: Similar metrics based on spectroscopic observations of the Sun with GOLF (black histogram) and HARPS-N (red histogram) compared with ESPRESSO observations of HD 67458 and HD 88595 (colored dashed and dotted vertical lines). Note that the relative scale of the radial velocity variations is significantly smaller compared to aniticipated oscillation amplitudes than is the case for the luminosity variations. Reprinted with permission from Sulis et al. (2023).

different sites. A third consideration is that, for classical pulsators and other, typically intermediate and high-mass stars, one analysis difficulty arises from *mode identification*. Successful modeling of oscillation spectra requires not simply knowing the frequencies, but also knowing to the greatest extent possible the quantum numbers *nlm* corresponding to each frequency. This is essentially because otherwise the universe of possible solutions is too large, making it impossible to select between possible model solutions. A number of different approaches are possible for mode identification, as discussed in Section 8.5, including photometric amplitude ratios in different wavelength regions and changes in the shape of individual line profiles caused by the oscillations. To date, no space photometry mission used for asteroseismology has had the ability to make simultaneous observations in different wavelength bands,[5] but high-resolution $R \sim 10^5$ ground-based observations are capable of resolving changes in line shape, enabling mode identification.

The amplitude of the radial velocity signal anticipated from oscillations in a solar-like star is of order 20 cm s^{-1}, implying that the inherent precision of any instrument must be considerably greater than this. For solar-like stars, resolving p-mode oscillations requires exposure times of a few minutes or less, which even for bright stars mandates relatively large telescopes (3m – class or more). Beyond shot noise, the main noise sources are stellar activity, granulation, and instrumental drifts. To limit the latter in particular, the instrument must be both intrinsically stable and provide a wavelength reference that is also precise and stable. Stability is primarily achieved through a combination of design features and environmental controls, focusing on limiting thermal drift, mechanical shocks, and pressure fluctuations. Typically the instrument is fiber-fed, isolated, and built into a vacuum chamber. While feeding the light through a fiber allows removal of the spectrograph from the intrinsically hostile environment at the telescope, it does not remove the problem that pointing drifts lead to uneven illumination of the fiber at the input, which can translate to exit beam drifts, which appear as small velocity changes; fiber scramblers are used to minimize this problem.

There are four basic approaches currently in use for providing a wavelength reference:

- The use of an iodine cell, which was proposed by Marcy & Butler (1992). In this technique, a transparent chamber filled with I_2 gas is placed in the spectrograph beam, and imprints its absorption spectrum onto the stellar spectrum. Iodine has a dense collection of strong narrow absorption lines at visible wavelengths, and the mass of the molecule limits thermal velocities. In theory, this is simple to implement, but it does have some complications. First, because they are in absorption, the signal-to-noise of the iodine lines is limited by the SNR of the stellar spectrum itself. Second, the complex spectrum of star + reference thatr results both complicates radial velocity analysis and makes ancillary science using the stellar spectra difficult or impossible as they are contaminated by the iodine spectrum. Finally, the analysis procedure requires templates of the spectra both of the iodine cell and of the uncontaminated stellar signal, the former at a resolution significantly higher than will be used, both of which add time and complexity. Nevertheless, the approach has had limited successes for asteroseismology, primarily for evolved stars and stars in the instability strip, which have higher radial velocity amplitudes.

[5] *Brite-Constellation* is an exception, though the fact that the different bandpasses here are observed using different satellites leads to calibration difficulties that can complicate analysis.

- Another approach involves using a ThAr or similar hollow-cathode lamp, illuminating the spectrograph simultaneously with the stellar source through a dedicated optical fiber. The lamp spectrum serves as a wavelength reference, and its SNR can be much higher than that of the stellar spectrum. Disadvantages of this approach include the fact that, unlike with the I_2 cell, the light path traversed by the calibration source is slightly different than that of the star. There is a wide variation in the brightness of ThAr lines, so some parts of the spectrum have few bright lines while others are so bright as to saturate the detector. Also, lamps evolve over time, with some lines fading and others appearing, and the overall brightness of the lamp changing as well, so consistent wavelength calibration can be challenging. Both iodine cells and ThAr lamps, when used alone, appear to be limited to somewhat better than 1 m s^{-1} precision.
- Fabry–Perot etalons consist of two parallel partially-reflecting plates, which produce a dense spectrum of transmission peaks that can be used as wavelength fiducials. They do require active control and precise calibration and stabilization of their components, but can be used in conjunction with other methods, such as hollow-cathode lamps, to reach combined precision an order of magnitude better than lamps on their own.
- Laser frequency combs produce a dense, evenly spaced collection of laser emission peaks that can be used as wavelength references. Together with high precision, they offer stability and an absolute wavelength reference standard when used together with a hollow-cathode lamp. Combs have reached precision levels of 1 cm s^{-1}, but have associated cost and maintenance challenges.

While for solar-like stars, SNR limitations due to short integrations can be challenging (though recent successes detecting oscillations in K dwarfs are promising), for O/B stars the challenge is the lack of available spectral lines, as well as their Doppler-broadened width in more rapidly-rotating stars.

Data reduction for fiber-fed echelle spectrographs follows a well-defined pathway (see, e.g., Piskunov et al. 2021), starting with bias subtraction and flat-fielding. The next step is identification and tracing of the curved spectral order paths across the detector, defining the diffuse scattered light inter-order background and removing it, and then extracting the spectra by summing the counts along each order, either in an unweighted or a more optimal way (Horne 1986). The resulting spectrum will have curvature due to the blaze and response functions of the dispersive elements, and typically that is handled by continuum normalization. The resulting spectrum is then wavelength-calibrated using a stable, well-characterized reference source: an iodine cell, hollow-cathode lamp, or laser frequency comb.

In the broadest sense, radial velocities are then determined by comparing spectra at different times to one another. Cross-correlation is one preferred method, and involves obtaining a mean-line profile of the stellar spectrum through cross-correlation with a mask, and obtaining the radial velocities by calculating changes in the profile centroid. The resulting cross-correlation profile is approximately

Gaussian in shape, and the radial velocity is obtained as the centroid of a functional fit to the cross-correlation function. Alternatively, a template spectrum can be constructed from all of the observed spectra, and using forward-modeling of the star + reference spectrum combined with a χ^2 or similar optimization approach to determine the shift of each individual spectrum relative to the template; this is typically the chosen approach when an iodine cell is used.

An important limitation on radial velocity measurements is that they rely on the presence of spectral lines, and on those lines being narrow; both of these lead to more steeply-sloped parts of the spectrum, which in turn produce stronger cross-correlation signals. Hot stars are therefore difficult, because they both have few lines and tend to be rapidly-rotating, leading to significant Doppler line broadening. At the cool end of the stellar effective temperature distribution, M star spectra are extremely complex and confusing, with overlapping molecular bands, with most of the limited flux being emitted in the infrared, where they are less accessible to ground-based instruments. Accordingly, the nature of the stars themselves tends to limit radial velocity measurements to the mid-F to mid-M range of spectral types on the main sequence; evolved stars are a bit more accessible due to their (usually) slower rotation. Cool stars with rapid rotation also tend to be chromospherically active, adding another intrinsic noise source along with stellar granulation noise.

In addition to the star, sources of radial velocity uncertainty include:

- Detector imperfections: CCD detectors suffer from flat-fielding uncertainties, variations in pixel size, thermal expansion and contraction, and non-zero charge transfer efficiency (CTE).
- Air: Changes in the index of refraction of air due to pressure and temperature changes are mapped to shifts in the wavelengths of spectral lines. Placing the spectrograph in an evacuated chamber minimizes these effects, which can amount to hundreds of m s^{-1} over a single observing night.
- Thermal variations: Mechanical and optical components in the spectrograph change size and flex during temperature changes, changing the focus and PSF of the spectrograph, changes which appear as velocity drifts. Thermal control of the spectrograph can limit these effects, though some portions of the optical train are always exposed.
- Fiber illumination variations: Fiber scramblers cause a lack of "memory" at the output end of the fiber about how it was illuminated, which limits the impact of telescope pointing jitter and atmospheric turbulence, but scrambling is always imperfect.
- Wavelength reference instability
- Telluric lines: Atmospheric lines appear in the spectrum and vary in width and depth over time, complicating cross-correlation.

Despite these concerns, the limiting precision of radial velocity measurements has increased with each generation of instruments. Currently, the best instruments (ESPRESSO, EXPRES, etc.) are essentially photon-limited (Figure 4.16), with noise sources intrinsic to the star such as granulation noise and stellar activity limiting rms radial velocity precision to around 0.5 m s^{-1} .

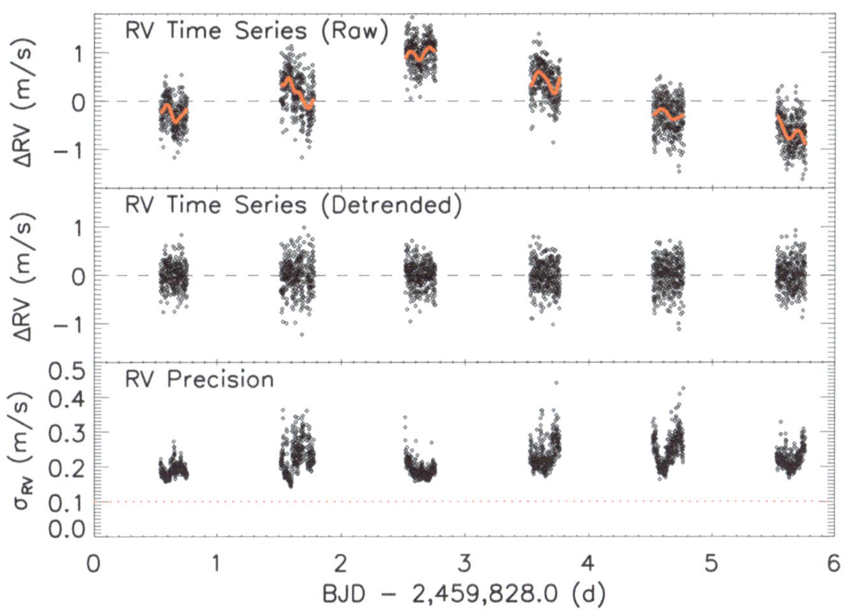

Figure 4.16. Intensity diagram of the residual flux after subtracting the mean line profile from each individual profile for a time series of the δ Scu star HD 41641. The result is then folded in phase with an oscillation period, making the periodic line profile variations readily apparent. Reprinted with permission from Campante et al. (2024).

References

Aigrain, S., Parviainen, H., & Pope, B. J. S. 2016, MNRAS, 459, 2408

Basu, S., & Chaplin, W. J. 2018, Asteroseismic Data Analysis. Foundations and Techniques (Princeton, NJ: Princeton Univ. Press)

Bedding, T. R., Butler, R. P., Kjeldsen, H., et al. 2001, ApJL, 549, L105

Bedding, T. R., Mosser, B., Huber, D., et al. 2011, Natur, 471, 608

Bertin, E., & Arnouts, S. 1996, A&AS, 117, 393

Bódi, A., Szabó, P., Plachy, E., Molnár, L., & Szabó, R. 2022, PASP, 134, 014503

Brown, T. M., Gilliland, R. L., Noyes, R. W., & Ramsey, L. W. 1991, ApJ, 368, 599

Buzasi, L., Carboneau, D., Hessler, L., Lezcano, C., & Preston, H., A. 2016, IAUFM, 29B, 673

Campante, T. L., Kjeldsen, H., Li, Y., et al. 2024, A&A, 683, L16

Christensen-Dalsgaard, J., & Frandsen, S. 1983, SoPh, 82, 469

Christensen-Dalsgaard, J., & Gough, D. O. 1982, MNRAS, 198, 141

Escorza, A., Zwintz, K., Tkachenko, A., et al. 2016, A&A, 588, A71

Feinstein, A. D., Montet, B. T., Foreman-Mackey, D., et al. 2019a, PASP, 131, 094502

Feinstein, A. D., Montet, B. T., Foreman-Mackey, D., et al. 2019b, Eleanor: Extracted and systematics-corrected light curves for TESS-observed stars, Astrophysics Source Code Library, ascl:1904.022

Frandsen, S., Fredslund Andersen, M., Brogaard, K., et al. 2018, A&A, 613, A53

Gibson, S. R., Howard, A. W., Marcy, G. W., et al. 2016, Proc. SPIE, 9908, 990870

Gilliland, R. L., & Brown, T. M. 1988, PASP, 100, 754

Gilliland, R. L., Chaplin, W. J., Dunham, E. W., et al. 2011, ApJS, 197, 6
Gizon, L., & Solanki, S. K. 2003, ApJ, 589, 1009
Halverson, S., Terrien, R., Mahadevan, S., et al. 2016, Proc. SPIE, 9908, 99086P
Handberg, R., Lund, M. N., White, T. R., et al. 2021, AJ, 162, 170
Horne, K. 1986, PASP, 98, 609
Huang, C. X., Penev, K., Hartman, J. D., et al. 2015, MNRAS, 454, 4159
Huang, C. X., Vanderburg, A., Pál, A., et al. 2020a, RNAAS, 4, 204
Huang, C. X., Vanderburg, A., Pál, A., et al. 2020b, RNAAS, 4, 206
Huang, C. X., Vanderburg, A., Pál, A., et al. 2020c, RNAAS, 4, 204
Jenkins, J. M., Caldwell, D. A., Chandrasekaran, H., et al. 2010, ApJL, 713, L87
Kjeldsen, H., & Bedding, T. R. 1995, A&A, 293, 87
Knudstrup, E., Lund, M. N., Fredslund Andersen, M., et al. 2023, A&A, 675, A197
Libralato, M., Bedin, L. R., Nardiello, D., & Piotto, G. 2016, MNRAS, 456, 1137
Lightkurve CollaborationCardoso, J. V. D. M., Hedges, C., et al. 2018, Lightkurve: Kepler and TESS Time Series Analysis in Python, Astrophysics Source Code Library, ascl:1812.013
Lund, M. N., Handberg, R., Davies, G. R., Chaplin, W. J., & Jones, C. D. 2015, ApJ, 806, 30
Lund, M. N., Handberg, R., Buzasi, D. L., et al. 2021, ApJS, 257, 53
Marcy, G., & Butler, P. 1992, PASP, 104, 270
Martínez-Palomera, J., Hedges, C., & Dotson, J. 2023, AJ, 166, 265
Nardiello, D., Borsato, L., Piotto, G., et al. 2019, MNRAS, 490, 3806
Nielsen, M. B., Ball, W. H., Standing, M. R., et al. 2020, A&A, 641, A25
Oelkers, R. J., & Stassun, K. G. 2018, AJ, 156, 132
Pepe, F., Cristiani, S., Rebolo, R., et al. 2021, A&A, 645, A96
Peri, M. L. 1995, PhD Thesis, California Institute of Technology
Petersburg, R. R., Ong, J. M. J., Zhao, L. L., et al. 2020, AJ, 159, 187
Petigura, E. A., Crossfield, I. J. M., Isaacson, H., et al. 2018, AJ, 155, 21
Piskunov, N., Wehrhahn, A., & Marquart, T. 2021, A&A, 646, A32
Plachy, E., Molnár, L., Bódi, A., et al. 2019, ApJS, 244, 32
Plachy, E., Pál, A., Bódi, A., et al. 2021, ApJS, 253, 11
Pope, B. J. S., White, T. R., Huber, D., et al. 2016, MNRAS, 455, L36
Provencal, J. L., Shipman, H. L., Montgomery, M. H., & Team, W. E. T. 2014, CoSka, 43, 524
Seifahrt, A., Stürmer, J., Bean, J. L., & Schwab, C. 2018, Proc. SPIE, 10702, 107026D
Stassun, K. G., Oelkers, R. J., Pepper, J., et al. 2018, AJ, 156, 102
Stassun, K. G., Oelkers, R. J., Paegert, M., et al. 2019, AJ, 158, 138
Stefansson, G., Li, Y., Mahadevan, S., et al. 2018a, AJ, 156, 266
Stefansson, G., Mahadevan, S., Wisniewski, J., et al. 2018b, Proc. SPIE, 10702, 1070250
Stumpe, M. C., Smith, J. C., Van Cleve, J. E., et al. 2012, PASP, 124, 985
Sulis, S., Lendl, M., Cegla, H. M., et al. 2023, A&A, 670, A24
Sullivan, P. W., Winn, J. N., Berta-Thompson, Z. K., et al. 2015, ApJ, 809, 77
Van Cleve, J. E., Howell, S. B., Smith, J. C., et al. 2016, PASP, 128, 075002
Vanderburg, A., & Johnson, J. A. 2014, PASP, 126, 948
White, T. R., Pope, B. J. S., Antoci, V., et al. 2017, MNRAS, 471, 2882
Woods, D. F., Ruprecht, J. D., Kotson, M. C., et al. 2021, PASP, 133, 014503

AAS | IOP Astronomy

Asteroseismology for the Nonspecialist

Derek L Buzasi

Chapter 5

Data Analysis Tools and Techniques

Once data have been processed to produce a light curve or a radial velocity curve, the next step is generally to search for the signatures of oscillations within those data. In some cases, when a star is oscillating with a large single amplitude relative to the noise level of the data, and when the time series is long enough to encompass many oscillations, both the presence of periodic variability and its characteristics—frequency, amplitude, and phase—are simple to derive. However, in most cases the oscillatory behavior is not readily obvious from simple inspection of the light curve. In those cases we make use of tools to translate the light curve (or velocity curve) from the time domain into the frequency domain.

In the simplest case we can consider a classical pulsator with a purely sinusoidal light curve observed in the very high signal-to-noise regime. Here, it is reasonable to simply perform a least-squares fit to the data and derive the frequency, amplitude, and phase information directly from that fit, along with uncertainties on these parameters. If the number of oscillation modes N is known, then one can extend this procedure further, to simultaneously fit multiple oscillation mode frequencies. However, in most real cases the number of modes is *not* known *a priori*, and the problem is thus more complicated.

One approach is to perform a least-squares fit to the data for a single mode, generally the one with the largest amplitude, and then make use of the derived parameters for that mode to remove it from the original time series, a process known[1] as *prewhitening*. The *residual* time series after prewhitening is then adopted as the new time series and the process is repeated until no statistically significant oscillation frequencies remain in the time series. This process or variants of it are frequently used in practice, though cases where individual frequencies are close to one another, or the underlying noise level is large or significantly non-white, can render it somewhat suspect. This can be remediated to some extent by using the

[1] Somewhat awkwardly.

doi:10.1088/2514-3433/ae03a0ch5

resulting list of frequencies, amplitudes, and phases as starting points for a multi-sinusoidal least-squares fit where the number of oscillation frequencies N used is set by the output from the prewhitening procedure.

However, for more complex cases, the fitting approach most commonly used is the *periodogram*, based on the *Fourier Transform*, which effectively allows the simultaneous fitting of a large number of test frequencies simultaneously (Lomb 1976; Scargle 1982; VanderPlas 2018). Prior to calculating the periodogram, some preparation of the light curve may be required, and large-amplitude oscillations which depart from sinusoidal shape, or oscillations whose amplitude is time-varying, may benefit from other methods, including autocorrelation functions (ACF; Roxburgh & Vorontsov 2006), wavelet transforms (Foster 1996; Ayres & Buzasi 2021; Pereira et al. 2024), phase dispersion modulation (PDM; Stellingwerf 1978; Schwarzenberg-Czerny 1997), Hilbert–Huang transform (Yan et al. 2024), short-time Fourier transforms (STFT), and others, but since Fourier analysis is by far the dominant technique used at present we refer the reader to the citations given.

5.1 Fourier Analysis

5.1.1 Intuitive Understanding

There is some pedagogical value in approaching the idea of a Fourier transform informally at first, so we will start that way. Let's start by imagining that we have two time series functions, f and g (both mean-subtracted and so centered about zero), and we want to characterize the *similarity* of these two functions, as illustrated in Figure 5.1. Looking ahead to the nature of the time series data we will need to use, we treat the functions as discretely sampled in time at some fixed interval δt. In this case, we can quantify the similarity between the two by taking a point-by-point multiplication of the two functions and summing it up: if the functions are similar to one another, the result will be a large positive value, while if the functions are

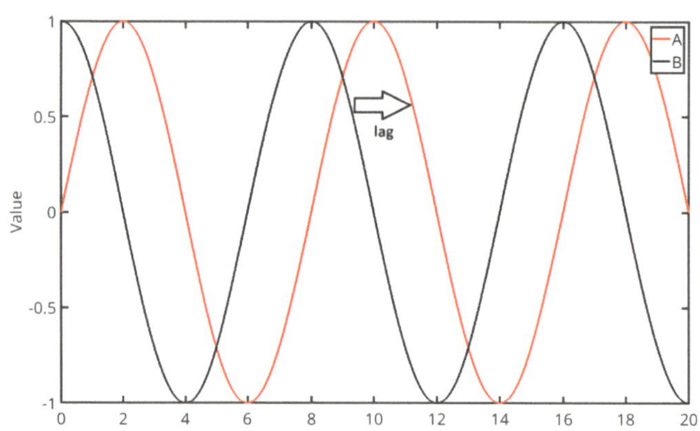

Figure 5.1. Two light curves or time series functions, A and B, discretely sampled in time and compared to one another. The lag parameter, as described in the text, is illustrated.

completely dissimilar the sum will result in a large negative value. If f and g are entirely uncorrelated, we can expect a sum close to zero.

$$f^{*}g = \sum_{n=1}^{N} f(t_n)g(t_n) \tag{5.1}$$

Of course, this doesn't take into account the fact that the function g shown in Figure 5.1 is really the same function as f, only offset in time, so we can see that the underlying functions really are similar. We can account for this by examining different *lags* τ for $g = g(t + \tau)$, as shown in Figure 5.1.

Now consider the idea of a *cross-correlation* between two functions. Looking ahead to the nature of the time series data we will need to use, we'll treat the functions as discretely sampled in time at some fixed interval δt, and we can define the cross-correlation between two functions f and g as

$$C_{fg} = f^{*}g \equiv \sum f(t)g(t - \tau) = \sum_{n=-\infty}^{\infty} f(t)g(t - n\Delta t) \tag{5.2}$$

A useful way to visualize this can be imagined from Figure 5.1, where we take the function g and slide it back and forth along the x axis, while keeping f fixed. In this context $\tau = t + n\Delta t$ is sometimes referred to as the *lag*. In practice, we never have infinitely long time series, so we only need to sum over the range of lags for which the two functions actually overlap.

Again, when both functions are large and have the same sign, the contribution to the summation is large and positive, while when both are large but have different signs, the contribution is large and negative. When the two are entirely uncorrelated, on average we will get zero. In general, a large positive value of C_{fg} means the two functions f and g are positively correlated.

Instead of g being some general function, next imagine that it is a sinusoid with some characteristic frequency ω, so $g(t) = \sin(\omega t - \tau)$, so

$$C_{fg}(\tau, \omega) = f^{*}g \equiv \sum f(t)\sin(\omega t - \tau) = \sum_{n=-\infty}^{\infty} f(t)\sin(\omega t - n\Delta t) \tag{5.3}$$

Essentially, here we are testing for the presence in the time series of a sinusoid at frequency ω; by varying the frequency ω, we can test for sinusoids of different frequencies. Clearly we can do the same for cosines using $\cos(\omega t + \tau)$, and we can use the complex representation

$$\exp(-i\omega t) = \cos(\omega t) - i\sin(\omega t) \tag{5.4}$$

implying

$$C_{fg}(\tau, \omega) = \sum_{n=1}^{N} f(t_n)\exp{-i\omega(t_n - \tau)} \tag{5.5}$$

Intuitively, such a convolution allows a means of testing simultaneously for the presence of a sine of frequency ω (the imaginary part of the convolution) and a cosine of frequency ω (the real part), which is the equivalent of testing for the presence of a sinusoid with frequency ω and a phase factor ϕ, where $\tan\phi = \mathrm{Im}(C_{fg})/\mathrm{Re}(C_{fg})$. The *power* of $C_{fg}(\omega)$ is $|C_{fg}(\omega)|^2$, and the (real) *amplitude* is $|C_{fg}(\omega)|$.

5.1.2 Formal Understanding

The *Fourier (Integral) Transform* offers a method of transformation between the time domain and the frequency domain of a signal. Formally, for a differentiable continuous finite-valued function, we can write it as

$$g(\omega) = \sqrt{\frac{1}{2\pi}} \int_{-\infty}^{\infty} f(t) e^{-i\omega t} dt \tag{5.6}$$

and its inverse as

$$f(t) = \sqrt{\frac{1}{2\pi}} \int_{-\infty}^{\infty} g(\omega) e^{i\omega t} d\omega \tag{5.7}$$

noting the difference in sign on the argument of the exponential within the integral between the forward and inverse transforms. It is important to be aware that different authors adopt differing conventions regarding the argument preceding the integral itself; while varying this choice does not affect the overall behavior of the transform, it clearly affects the specific values derived, so it is good practice to verify the convention being used by any given author.

One important result concerning the Fourier transform is *Parseval's Theorem*, which states that

$$\int_{\infty}^{\infty} |f(t)|^2 dt = \int_{\infty}^{\infty} |g(\omega)|^2 d\omega \tag{5.8}$$

In other words, the total power in the time domain is equal to the total power measured in the frequency domain. This can be of practical use in scaling the output of a power spectrum estimation, as it is approximately true even in real circumstances when the transform is discrete and the limits no longer range over all times and frequencies.

Another important relationship can be illustrated by taking the transform of a single rectangular pulse, as in the left side of Figure 5.2. Here $f(t) = 1$ for $|t| < c$ and zero otherwise. In this case, we can calculate the transform analytically, so that

$$g(\omega) = \sqrt{\frac{1}{2\pi}} \int_{-\infty}^{\infty} f(t) e^{-i\omega t} dt = \sqrt{\frac{1}{2\pi}} \int_{-c}^{c} e^{-i\omega t} dt = \frac{e^{-i\omega t}}{-i\omega}\Bigg|_{-c}^{c} = \frac{e^{i\omega c} - e^{-i\omega c}}{i\omega} \tag{5.9}$$

Making use of the relation

$$\frac{e^{ix} - e^{-ix}}{i\omega} = 2\sin x \tag{5.10}$$

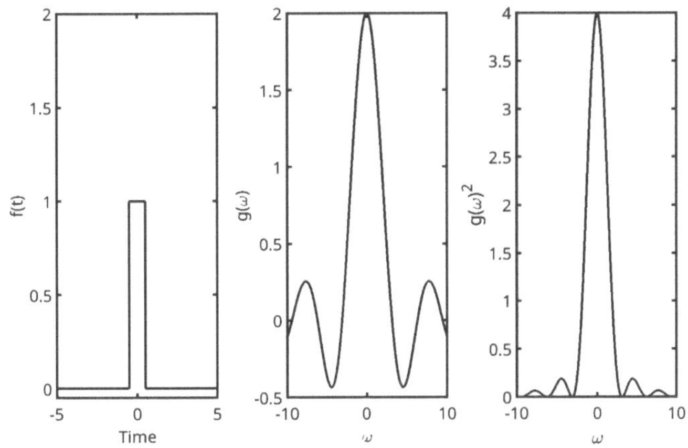

Figure 5.2. Left: A rectangular pulse of amplitude and length unity, centered on $t = 0$. Center: The amplitude spectrum of the pulse on the left. Right: The power spectrum of the pulse.

allows us to write

$$g(\omega) = \frac{2 \sin \omega c}{\omega} \tag{5.11}$$

a real-valued function which is shown in the center of Figure 5.2. We can determine the value of this function at $\omega = 0$ as

$$2\lim_{\omega \to 0} \frac{\sin \omega c}{\omega} = 2c \lim_{x \to 0} \frac{\sin x}{x} = 2c \tag{5.12}$$

adopting $x \equiv \omega c$ and recalling that

$$\lim_{x \to 0} \frac{\sin x}{x} = 1 \tag{5.13}$$

The first zeros of the function are trivially $\pm\pi/c$. Thus, as the length of the pulse c increases, the energy of the pulse is concentrated into a narrower and narrower window in the frequency domain. Even more generally, since $c = \Delta t$ is the width of the pulse in the time domain,

$$\Delta\omega\Delta t = \left(\frac{\pi}{c}\right)c = \pi \tag{5.14}$$

The width of the pulse in the time domain and the shape of its amplitude spectrum (or power spectrum, shown in the right-hand panel of Figure 5.2) are related to one another. In the most extreme case, where the pulse width in the time domain becomes infinite and $c \to \infty$, the amplitude spectrum becomes a delta function located at the origin, with effectively zero bandwidth.

The previous section implied the existence of a relationship between the Fourier transform and the convolution, and we can make that relationship formal through the Convolution Theorem, which states that the Fourier transform of a convolution is the product of the individual Fourier transforms, with the product in this case being taken on a point-to-point basis, so

$$\mathscr{F}\{f^{*}g\} = \mathscr{F}\{f\} \cdot \mathscr{F}\{g\} \tag{5.15}$$

The overall near-symmetry of the Fourier transform suggests, correctly, that the reverse is also true, so

$$\mathscr{F}\{f \cdot g\} = \mathscr{F}\{f\}^{*}\mathscr{F}\{g\} \tag{5.16}$$

In practice, our time series are discretely sampled rather than continuous, so we use the *Discrete Fourier Transform* (DFT), which takes the form

$$g(\omega) = \frac{2}{N}\sum_{j=1}^{N} f(t_j)e^{-i\omega t_j} \tag{5.17}$$

where we have a time series of N points. In practice the majority of practitioners work with linear frequency ν rather than angular frequency $\omega = 2\pi\nu$, making this

$$g(\nu) = \frac{2}{N}\sum_{j=1}^{N} f(t_j)e^{-2\pi i\nu t_j} \tag{5.18}$$

One way to understand the movement from the continuous to the discrete case is as a transform of the pointwise product of the signal and another function, the observing window, so

$$f_{\rm obs}(t) = f(t) \cdot W(t) \tag{5.19}$$

In the idealized case where our observations are infinitely short, the window function is just a collection of delta functions corresponding to the observing times,

$$W(t) = \sum_{i=1}^{N} \delta(t - t_i) \tag{5.20}$$

When we take the Fourier transform, we then produce a convolution of the transforms of the signal and the window function, so

$$g_{\rm obs} = g(\omega)^{*}g(W) \tag{5.21}$$

Since the window function is discrete, this produces a discrete transform. If our observations have length δt and have no gaps between them, then we are effectively producing a Fourier transform convolved with a sinc function (Equation (5.11)) with width $1/(N\delta t)$.

As in the continuous case, we can think of the Fourier Transform as a map from the time to the frequency domain, and vice versa. The information contained in both descriptions is the same, and the transform contains information about both the *amplitude* and the *phase* at each frequency. In general, the transform is a complex--valued function, and phase information can be derived from the ratio of the imaginary to the real components in the usual way

$$\tan[\phi(\omega)] = \frac{\mathrm{Im}(g(\omega))}{\mathrm{Re}(g(\omega))} \tag{5.22}$$

Of course the absolute value of the phase is dependent on the initial time adopted for $t = 0$, so normally we discuss only relative phases, and in many analyses the phase is ignored entirely. However, there are circumstances under which it encodes valuable physical information.

The power at each frequency $P(\nu)$ is proportional to the squared magnitude of the complex-valued transform, so that

$$P(\nu) \sim A^*A = |A(\nu)|^2 \tag{5.23}$$

As the number of points becomes large (greater than 100 or so in practice), the constant of proportionality approaches $N/4$, so we can write

$$P(\nu) = \left(\frac{4}{N}\right)|A(\nu)|^2 \tag{5.24}$$

A plot of $P(\nu)$ as a function of frequency is the *power spectrum*, and Parseval's theorem can also be used to properly scale the power spectrum. Alternatively, we can use the *amplitude spectrum*, which is just the square root of the (real) power spectrum. Authors vary as to whether to show amplitude or power spectra. Power spectra tend to make the quality of the data look better because they emphasize the high *SNR* peaks, while one can argue that amplitude spectra are more immediately related to the physical quantity of most interest: the amplitude of each pulsational mode.

Another note: formally, what we've been referring to as the *power spectrum* is really the *periodogram*. The power spectrum is a continuous function and the periodogram is a discrete *estimator* of that function. However, like most astronomers, we will continue with the inexact usage of using the two terms roughly equivalently, reminding the reader that real data are in fact discrete.

5.1.3 Resolution Limitations of Real Data

An important aspect of the amplitude (or power) spectrum in common use is its frequency resolution, or our ability to separate two closely-spaced frequencies. We've already seen in Equation (5.11) that the shape of the amplitude spectrum for a finite pulse in the time domain is just

$$g(\omega) = \frac{2\sin\omega c}{\omega} = 2c\,\mathrm{sinh}\,\omega c \tag{5.25}$$

which has a characteristic width $\sim 1/c$. A more realistic simple case would involve a purely sinusoidal variability over a limited time T. Note that the finite time interval could represent a limited observation interval (we rarely get to observe targets for infinite time!), or it could simply mean that the lifetime of the oscillation in question is T: both produce equivalent results. In either case, the transform of the finite sinusoid is also a sinh function centered on the sinusoidal frequency ω_0, with characteristic width $\sim 1/T$. One might reasonably conclude that this represents the frequency resolution available using from a time series of length T, and it *is* a reasonable first estimate. However, we have not considered the effects of noise, gaps, or uneven spacing in the time series, aliasing, or other departures from single-oscillator infinite signal-to-noise case. These we will explore later, in Sections 5.2, 5.3, and 5.4.

5.2 Aliasing and Nyquist Sampling

Let's look at the power spectrum for a simple sinusoid with amplitude unity and a period of 1 day and examine the effects of *sampling*, the cadence with which we collect data. For now, we'll assume the cadence is fixed, and later we can relax that assumption. Further assume that the phase $\phi = 0$, so our input signal is

$$F = \sin(2\pi f_s t + \phi) = \sin(2\pi t) \tag{5.26}$$

This is the red curve shown in Figure 5.2. If we now *sample* the sinusoid at intervals of 0.75d, and calculate the amplitude spectrum from $f = 0$ to $f = 2\ \mathrm{d}^{-1}$ in steps of $0.01\ \mathrm{d}^{-1}$, scaling as shown in Section 5.1.2, we obtain the result shown in Figure 5.4. Clearly we detect the original frequency at $1\ \mathrm{d}^{-1}$, but there's an additional peak visible at $0.333\ \mathrm{d}^{-1}$, which we note has a frequency equal to the *difference* 1.333–$1.000 = 0.333\ \mathrm{d}^{-1}$ between the signal and sampling frequencies. To see why, let's plot time series for both of these frequencies on top of one another, as shown in the black curve shown in Figure 5.3. Notice that the two curves cross one another, and the crossing points are marked with the open circles. By simple inspection, it's clear that the crossing points are separated by 1 d: if we sample these curves at a cadence of 0.75 d, they look *exactly* the same, and the power spectrum simply reflects that reality, though the exact crossing point locations of course depend on the relative phases of the two curves! This phenomenon is called *aliasing*.

Aliasing means that there is a relation between the sampling frequency f_s and the highest unambiguously detectable frequency, known as the Nyquist frequency

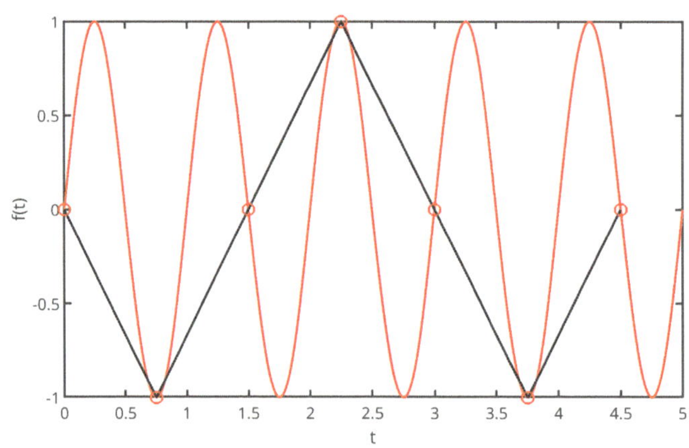

Figure 5.3. In red, a sinusoidal light curve with amplitude unity and a period of one day. The black curve shows the effects of sampling the signal with a cadence of 0.75d.

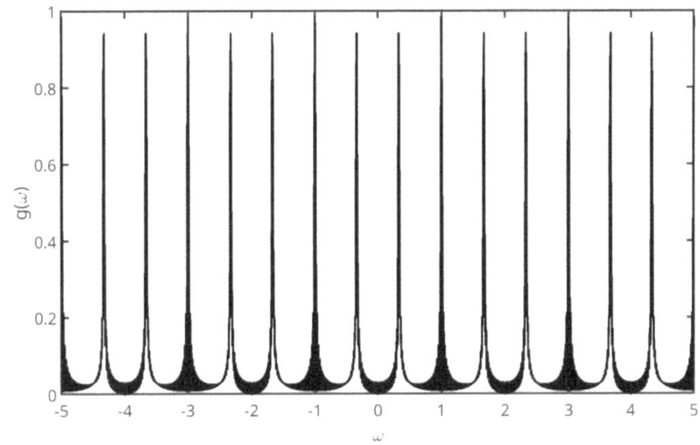

Figure 5.4. The amplitude spectrum of the black curve shown in Figure 5.2. Due to undersampling, the true peak at $1\mathrm{d}^{-1}$ is accompanied by alias peaks at higher and lower frequency.

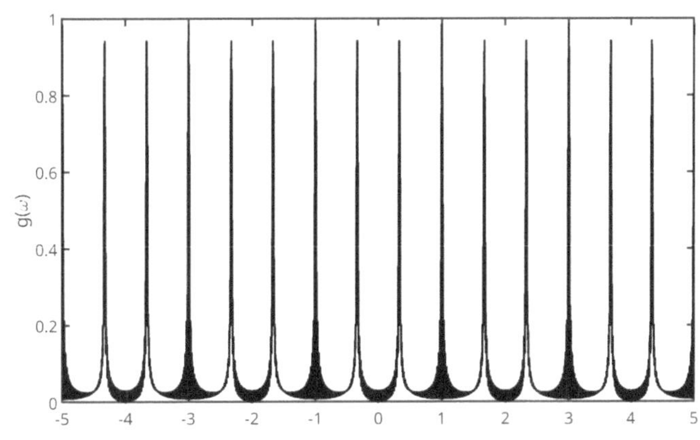

Figure 5.5. Same as Figure 5.4, but calculated for a wider range of both positive and negative frequencies, to show the extended pattern of aliases caused by undersampling.

$$f_{\mathrm{Ny}} = \frac{f_s}{2} \tag{5.27}$$

In other words, we need to sample each periodic component at least twice during a cycle. In the case shown above, our sampling frequency of 0.75 d^{-1} implies a Nyquist frequency of 0.375 d^{-1}; since our input frequency was above this, we experience aliasing due to *undersampling*. Note that showing a wider range of frequencies over which we calculated the amplitude spectrum (Figure 5.5) shows that the pattern of aliased peaks repeats itself indefinitely with increasing frequency. Furthermore, while an alias peak appears at $f - f_s = 0.333\ \mathrm{d}^{-1}$, another appears at $f + f_s = 1.667\ \mathrm{d}^{-1}$.

Extending the range of frequencies to negative values can also help to intuitively understand the third peak in Figure 5.4, at 1.667 d^{-1}: peaks reflect about the origin, so the original (correct!) peak at $1.000\,\mathrm{d}^{-1}$ is aliased to peaks at $0.333\,\mathrm{d}^{-1}$, $-0.333\,\mathrm{d}^{-1}$, $-1.000\,\mathrm{d}^{-1}$, and $-1.667\,\mathrm{d}^{-1}$, and that last peak is mirrored about the origin to appear at $+1.667\,\mathrm{d}^{-1}$. In practice, aliasing is considerably more of a nuisance even than this, because we normally have multiple frequencies present, each with its own pattern of alias peaks overlaid on one another. The presence of noise also poses a problem, because it can make alias peaks appear with larger amplitudes than the true oscillation frequency peak.

How do we address this problem? One approach is to increase the cadence of the original time series. Figure 5.6 shows the impact of increasing the cadence by a factor of ten, which removes the alias structure entirely. This approach was taken with the *TESS* mission: initially short-cadence data were sampled every two minutes, but later in the mission this was reduced to 20 seconds. The corresponding Nyquist frequencies are 4.167 and 25 mHz.

The Nyquist frequency defines the maximum unambiguously detectable frequency in the time series (with some provisos as noted below), but there is also a *minimum* detectable frequency, which is governed by the total length of the time series. This makes intuitive sense because as the period of variation grows to approximate and then exceed the length of the time series, the time series samples a smaller and smaller portion of the period and thus becomes a poorer and poorer representation of the underlying data. Figure 5.7 illustrates this effect, by showing the impact of decreasing the frequency of the input sinusoid from 0.33 to $0.033\,\mathrm{d}^{-1}$. Note that as the frequency drops below about $1/T$, where T represents the temporal span of the time series, the ability to determine reliably both the frequency and the amplitude of the input sinusoid vanish.

We can now come back to discuss an issue that we've glossed over so far. When we write the DFT as

$$g(\nu) = \frac{2}{N}\sum_{j=1}^{N} f(t_j)e^{-2\pi i\nu t_j}, \tag{5.28}$$

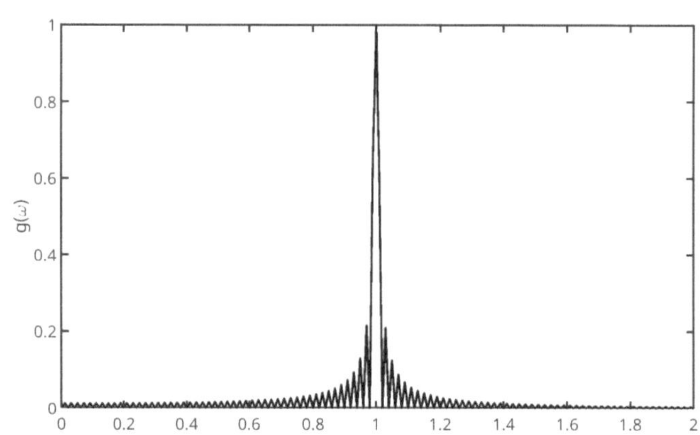

Figure 5.6. Same as Figure 5.4, but with the sampling cadence increased by a factor of 10 to show the removal of the alias pattern.

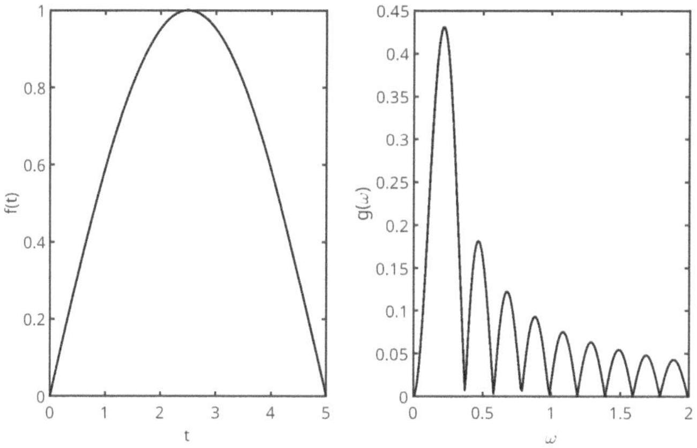

Figure 5.7. Left: An input sinusoid, which is undersampled, so that a complete cycle does not occur during the observation period. Right: The amplitude spectrum of the time series shown to the left. The input time series has been mean-subtracted to remove what would otherwise be a dominant peak at zero frequency. Note that both period and amplitude are incorrect.

we've said nothing about exactly which frequencies ν at which to evaluate the DFT! While they need not be evenly spaced, so that $\nu_j = \nu_0 + j\delta\nu$, computational ease of implementation and efficiency mean they generally are, and usually $\nu_0 = 0$, though it need not be. In that case, we'd expect that $\delta\nu \sim 1/T$ as suggested by the discussion of the frequency resolution in Section 5.1.3, and the application of Nyquist sampling would suggest that we should sample at twice that cadence, so

$$\delta\nu = \frac{1}{2T} \tag{5.29}$$

In practice, with real data, frequency sampling at this cadence risks either missing the tops of narrow peaks and underestimating their amplitudes, or even missing them entirely, so frequency spectra are generally oversampled at least 4 or 8 times the minimum rate.[2] For input data evenly spaced in the time domain by some Δt, the upper limit for calculating $g(\nu)$ derives directly from the Nyquist frequency and is $\nu = \frac{1}{2\Delta t}$.

5.3 Noise Sources and Effects

Real data always come with added noise. The most fundamental limiting noise for photometry is of course photon or shot noise, which is Poisson in nature and scales like $\sim 1/\sqrt{N}$, where N is the total number of photons counted in each observation. Photon noise can be driven to arbitrarily low levels by increasing the telescope aperture (with concomitant cost increases) or increasing the length of the exposure (which simultaneously decreases the cadence and frequency resolution possible). Perhaps most importantly for asteroseismic analysis, however, photon noise is *white*,

[2] VanderPlas (2018) argues for a value between 5 and 10

by which we mean that its level is frequency-independent. More formally, its integrated amplitude is the same over any bandpass in the amplitude spectrum; visually, this means that the power spectrum of such noise is basically flat (Figure 5.8). Note that this is a log-log plot, which helps show the frequency-independent nature of white noise (and the not-so-independent nature of other kinds; see below!) Another way of understanding white noise is that it has no time dependence, so each point in the time series is uncorrelated with any other point.

Most other types of noise are time-correlated in some way, and thus are *not* white. Many of these manifest themselves in the power spectrum such that the noise level rises at lower frequencies, generally following

$$P \sim \nu^{-\beta} \tag{5.30}$$

In this formulation white noise has $\beta = 0$, and we speak of "red noise" ($\beta = 2$) and "pink noise" ($\beta = 1$). Sometimes the terminology $1/f$ noise is used, but this can be confusing because noise can be inversely proportional to frequency measured in amplitude ("red noise") or power ("pink noise"). Figure 5.8 illustrates both white (black line) and red (red line) noise. Note that both are scaled to have identical variances σ^2.

The different types of noise have important impacts on how one chooses to both take data and analyze it. In the time domain, the different types of noise look different from one another, as can be seen in Figure 5.9. Once again, the black line indicates white noise, while the red line corresponds to red noise. Both have the same variance, but the "memory" inherent in red noise is apparent and leads to this kind of noise sometimes being called *drift noise*. In the white noise case, as noted above, increasing the sampling or integration time decreases the noise per data point, as $\sqrt{N}$, where N is the number of points. For red noise, however, increasing the integration time (or, equivalently, binning your data) will actually *increase* the noise per data point. In the pink noise case, the scatter of data points will be independent of whether integrations are short and many in number, or longer and fewer.

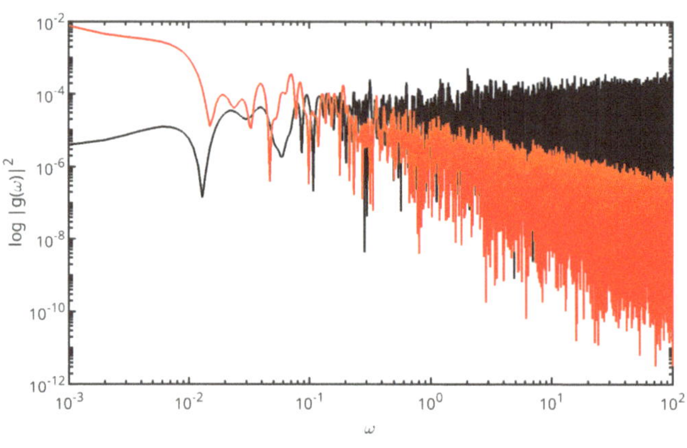

Figure 5.8. A log-log noise plot of the power spectrum of white noise (black curve) and red noise (red curve). Both input time series have been scaled to have identical variances.

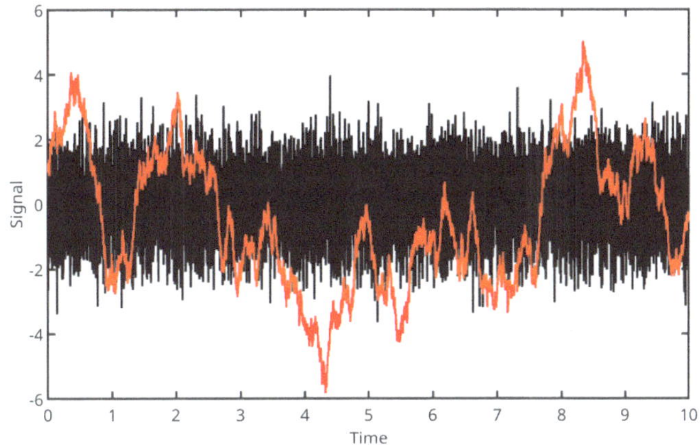

Figure 5.9. The corresponding time series from Figure 5.8, showing the white noise (black curve) and red noise (red curve) time series. The "drift noise" behavior of red noise is apparent.

Specific generators of noise vary widely, and in some cases interpretations of what actually constitutes "noise" do as well.[3] For space-based asteroseismology missions such as Kepler and TESS, important contributors include:

- Spacecraft pointing jitter. This moves the stellar image on the detector, creating a relatively complex mapping of the image motion plus detector pixel and sub-pixel response variations onto the measured stellar flux. Imperfections in flat-fielding response structure across the detector also contribute.
- Detector imperfections. These include CCD readout noise ("read noise"), non-linear response, accumulated radiation damage, and amplifier drifts (typically due to temperature variations).
- Spacecraft variations. In addition to attitude instabilities (manifested as pointing jitter), spacecraft flexure and changes in telescope focus, both magnified by thermal changes due to heating from both internal and external sources.
- Scattered light, particularly as it varies with time, and when due to bright sources moving in and out of the field, or scattering light into the field of view from nearby.
- Timing: Imperfections or irregularities in the spacecraft clock can manifest as increased noise in the light curve.
- Stellar: Stellar activity and granulation both produce time-varying aperiodic (or quasi-periodic) signals in both velocity and luminosity.

5.4 Gaps and Uneven Sampling

When data are evenly spaced, calculating the Fourier transform and the periodogram or power spectrum are greatly accelerated by the Fast Fourier Transform algorithm. In astronomy in general and asteroseismology in particular, however,

[3] One person's data is another person's noise.

ground-based astronomers don't usually get to choose their observing times to be perfectly regular, and even with space-based photometry, where observations are (to first order at least) intended to be evenly spaced, data dropouts, loss of guidance, breaks to enable telemetry, cosmic ray events, and a host of other factors produce data which are in practice irregularly spaced, or at least gapped. What effect does this have?

The window function discussed above for evenly spaced data in the time domain leads to evenly spaced peaks in the frequency domain, in the power spectrum. The typical structure is visible in Figure 5.6, which shows a main peak (the true frequency) surrounded by a small forest of peaks whose amplitude declines as one moves away from the main peak and whose spacing is approximately $1/T$: this is the window function in the frequency domain. Introducing gaps in the time series introduces complexity to the window function, which in turn produces complex alias structure, changing the amplitudes of peaks and introducing new ones. Consider simply breaking the time series into two unequal chunks, where now each introduces its own (generally different and interacting) window function to the power spectrum. More unevenness in the time domain produces a more complex window function in the frequency domain, which essentially introduces noise to the power spectrum. Figure 5.10 shows the result: the same time series as in Figure 5.6, but with half the regularly-spaced observations randomly removed.

An advantage that uneven sampling can offer, however, is regarding the Nyquist frequency. As the time series departs more and more from being randomly sampled, the Nyquist frequency becomes less and less meaningful, and frequencies above the traditional $\nu_{\rm Nyq}$ become accessible. In the end, the new effective Nyquist frequency can be as high as $1/2t_{\rm obs}$, where $t_{\rm obs}$ is the length of an observation, and this can potentially be orders of magnitude higher than the Nyquist frequency for the power spectrum of an evenly spaced time series. Of course, computation time can become

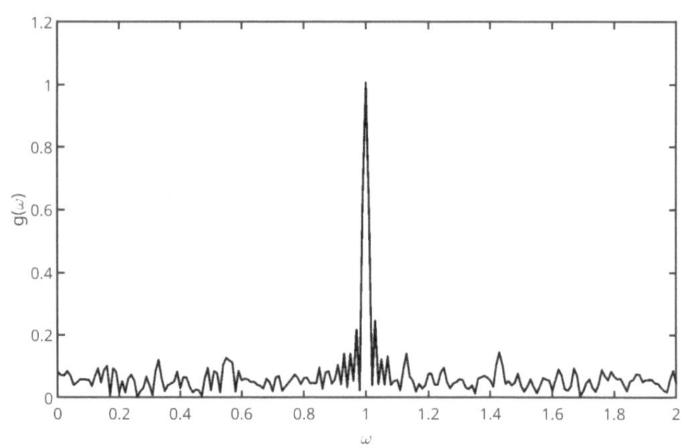

Figure 5.10. The amplitude spectrum of the same noiseless time series shown in Figure 5.6, but with half the points randomly removed to create an unevenly-spaced time series. Note the increased noise in the amplitude spectrum away from the main peak.

excessive in that case, so generally the upper frequency limit is chosen guided by prior knowledge of frequencies expected to be present.

In some cases, quasi-irregular sampling can be used to perform *super-Nyquist* frequency detection (Murphy et al. 2013; Murphy 2015; Greiveldinger et al. 2023). In the case of Kepler, the spacecraft orbit necessitates barycentric corrections to the observation timestamps, which significantly (and periodically) break the otherwise-regular sampling intervals. This breaks Nyquist alias peaks into multiplets, which can then be distinguished from the "real" peaks, which are not so affected. Similar approaches are likely possible for other satellite missions as well, or could be built into the operating modes for the mission (Murphy 2015; Shibahashi & Murphy 2018), though the approach is not foolproof and should be handled with care (Skarka et al. 2022).

An approach sometimes taken when gaps in the time series are short is simply filling the gaps, sometimes referred to as "infilling" or "inpainting" (García et al. 2014). The process used varies from simply replacing the missing points with a local mean value or a polynomial or spline interpolation, to estimating missing values using a Gaussian process, to effectively resampling the time series onto a uniform grid using a discrete cosine transform (García et al. 2014). Proponents argue that the approach lowers the overall background noise level, and can have significant impacts on the estimation of asteroseismic quantities such as $\nu_{\max}$. However, others argue that gap-filling produces artificial structures in the power spectrum, which themselves can affect measurements of oscillation frequencies and linewidths (and hence mode lifetimes) (Bedding & Kjeldsen 2022). Conservative analysis would suggest using any gap-filling procedure sparingly.

5.5 Lomb–Scargle Periodogram

An alternative to the DFT is the Lomb–Scargle Periodogram (Scargle 1982; Schwarzenberg-Czerny 1998; VanderPlas 2018), defined as

$$P_{LS}(\nu) = \frac{1}{2}\Bigg[\left(\sum_n g_n \cos(2\pi\nu[t_n - \tau])\right)^2 \Big/ \sum_n \cos^2(2\pi\nu[t_n - \tau]) + \left(\sum_n g_n \sin(2\pi\nu[t_n - \tau])\right)^2 \Big/ \sum_n \sin^2(2\pi\nu[t_n - \tau])\Bigg] \quad (5.31)$$

where τ is given by

$$\tau = \frac{1}{4\pi\nu}\tan^{-1}\left[\frac{\sum_n \sin(4\pi\nu t_n)}{\sum_n \cos(4\pi\nu t_n)}\right] \quad (5.32)$$

The similarities and differences between the LS periodogram and the classical are made clearer if we write the latter in a similar form,

$$P(\nu) = \frac{1}{N}\left[\left(\sum_n g_n \cos(2\pi\nu t_n)\right)^2 + \left(\sum_n g_n \sin(2\pi\nu t_n)\right)^2\right] \tag{5.33}$$

The changes are primarily in the scaling, and in the statistical properties of the periodogram when the time series in unevenly sampled. In that case, the Lomb–Scargle periodogram produces χ^2 distributed values when applied to white noise, while the classical periodogram only does so for evenly-sampled data.

Interestingly, the Lomb–Scargle periodogram can be interpreted as the result one would get by fitting a single sinusoidal model of the form

$$y(\nu) = A_\nu \sin[2\pi\nu(t - \phi_\nu)] \tag{5.34}$$

to the time series at every frequency, and constructing $P_{\mathrm{LS}}(\nu)$ as the χ^2 goodness-of-fit metric at each such frequency. A fairly common variant of the periodogram is effectively the modification of the model to include an offset term,

$$y(\nu) = y_0(\nu) + A_\nu \sin[2\pi\nu(t - \phi_\nu)] \tag{5.35}$$

which improves the solution when, for example, frequencies are not fully sampled over phase. This variant is sometimes called the *generalized* or *floating-mean Lomb–Scargle* periodogram.

Perhaps the strongest selling point of the Lomb–Scargle periodogram is that its statistical properties lead to a straightforward way to estimate the *false alarm probability* (FAP) at any frequency,

$$Prob(P) = 1 - \exp(-P) \tag{5.36}$$

where P is the periodogram value at that frequency. This is the (cumulative) probability of observing a periodogram value less than Z from a time series consisting only of white noise. However, this is **not** the probability of seeing a single peak greater than Z in the entire periodogram, which is the false alarm probability that really interests us! That probability is much harder to calculate, primarily because the periodogram value at some frequency $P_{\mathrm{LS}}(\nu)$ is not independent of the value of the periodogram at other—particularly neighboring—points! We discuss this situation in more detail below, in Section 7.2.

The Lomb–Scargle periodogram can be calculated quickly, because modern implementations cleverly make use of the FFT algorithm (Leroy 2012), and on the face of it has improved statistical quantities for unevenly spaced data and a straightforward algorithm for calculating the FAP. However, in real data, the assumption of white noise behind the improved statistical properties is rarely correct, particularly for low-frequency peaks, and the assumption that the model signal is sinusoidal may not be right either, particularly for high-amplitude peaks. In addition, while the height of a peak in the classical periodogram is clearly related directly to the amplitude of a sine wave at that frequency, the height of such a peak in the Lomb–Scargle periodogram is not so intuitive, particularly for very unevenly sampled data. Both classical and Lomb–Scargle periodograms are often seen in the literature.

References

Ayres, T., & Buzasi, D. 2021, RNAAS, 5, 243

Bedding, T. R., & Kjeldsen, H. 2022, RNAAS, 6, 202

Foster, G. 1996, AJ, 112, 1709

García, R. A., Mathur, S., Pires, S., et al. 2014, A&A, 568, A10

Greiveldinger, A., Garnavich, P., Littlefield, C., et al. 2023, ApJ, 955, 150

Leroy, B. 2012, A&A, 545, A50

Lomb, N. R. 1976, ApS&S, 39, 447

Murphy, S. J. 2015, MNRAS, 453, 2569

Murphy, S. J., Shibahashi, H., & Kurtz, D. W. 2013, MNRAS, 430, 2986

Pereira, A. W., Janot-Pacheco, E., Emilio, M., et al. 2024, A&A, 686, A20

Roxburgh, I. W., & Vorontsov, S. V. 2006, MNRAS, 369, 1491

Scargle, J. D. 1982, ApJ, 263, 835

Schwarzenberg-Czerny, A. 1997, ApJ, 489, 941

Schwarzenberg-Czerny, A. 1998, MNRAS, 301, 831

Shibahashi, H., & Murphy, S. J. 2018, PHysics of Oscillating STars. Proceedings from the PHOST (PHysics of Oscillating STars) Symp. (Arras: Sciencesconf) 22

Skarka, M., Žák, J., Fedurco, M., et al. 2022, A&A, 666, A142

Stellingwerf, R. F. 1978, ApJ, 224, 953

VanderPlas, J. T. 2018, ApJS, 236, 16

Yan, S.-P., Ji, L., Zhang, P., et al. 2024, RASTI, 3, 56

Derek L Buzasi

Chapter 6

From Frequencies to Physics: Basic Topics

6.1 Measuring the Power Spectrum

6.1.1 Solar-Like Main Sequence Stars

The Sun, the star we know best, can serve as an exemplar of lower main sequence stars, and its power spectrum is shown in Figure 6.1, based on data from the Virgo experiment. The main large-scale features are the peaks that are approximately regularly-spaced in frequency, and the Gaussian-shaped envelope defining the peak heights. The peak at the center of the envelope is the frequency of maximum power ν_{max}, and the average separation between the peaks is $\Delta\nu/2$.[1] Notice that the overall amplitude of the oscillations is larger in the blue than in the red.

A closeup view of the power spectrum, as in Figure 6.2, shows a somewhat more complicated structure: the large nearly regularly-spaced peaks are each split into pairs. We can define different large separations, $\Delta\nu_l$, representing separations between oscillation modes with the same radial mode n but differing degree l. We can also define *small separations* such that the average small separation is given by

$$\delta\nu \approx (4l + 6)D_0 \tag{6.1}$$

where as we saw in Section 3.6.3

$$D_0 = -\frac{\Delta\nu}{4\pi^2\nu_{nl}}\int_0^R \frac{dc}{dr}\frac{dr}{r} \tag{6.2}$$

Most of the contribution to the integral D_0 arises near the center of the star due to the $1/r$ term, so it is a valuable diagnostic for the deep interior. In fact since

$$c^2 = \frac{\gamma P}{\rho} \approx \frac{\gamma k T}{\mu m_{\text{h}}}, \tag{6.3}$$

[1] The factor of 2 arises because the large separation is the separation between successive values of n with the same l. See Equation (3.145).

doi:10.1088/2514-3433/ae03a0ch6

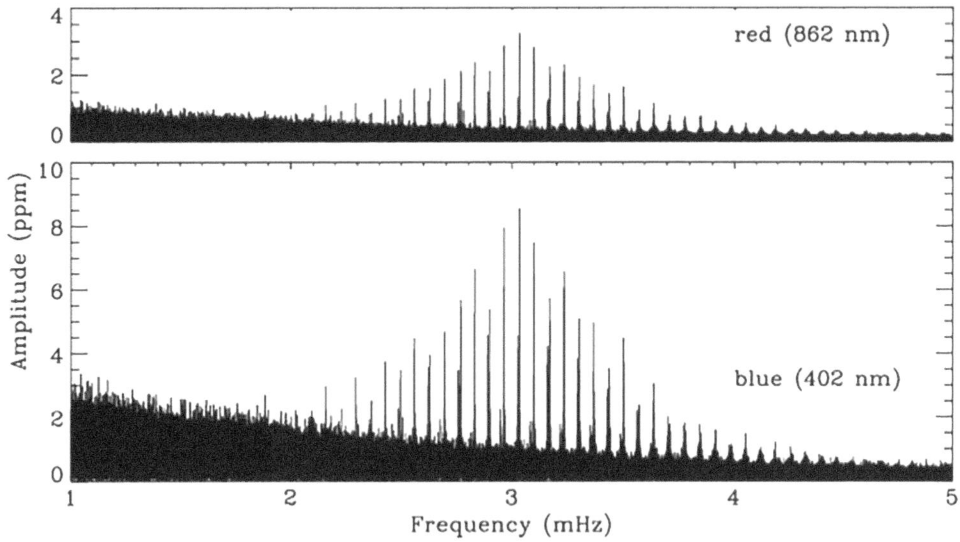

Figure 6.1. The amplitude spectrum of solar oscillations, measured using the VIRGO instrument on the *SOHO* satellite. The frequencies are smoothed to correspond to an observing time of 30 days, for better comparison to stellar data. Reprinted with permission from Bedding & Kjeldsen (2003).

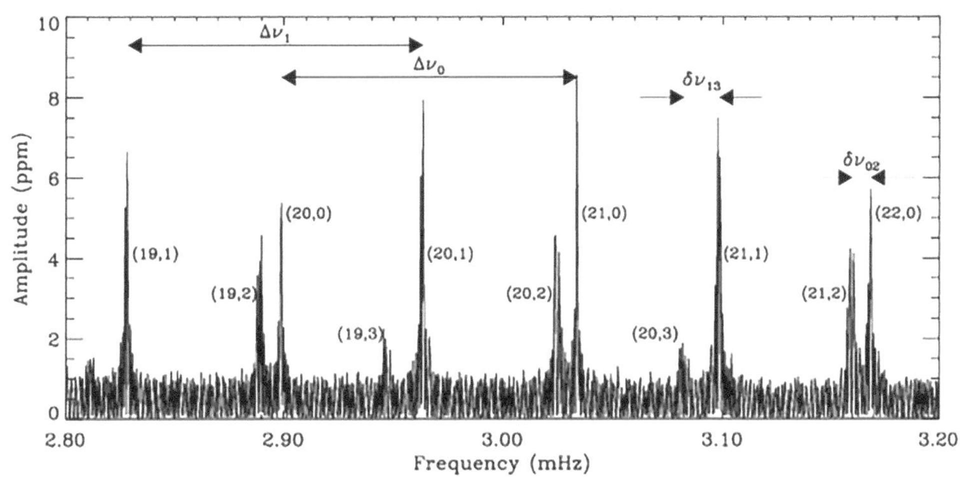

Figure 6.2. A small section of the (blue) solar amplitude spectrum from Figure 6.1, showing n, l values for each mode and labeling the various large and small frequency separations. Reprinted with permission from Bedding & Kjeldsen (2003).

we can pin some secular changes to the age of the star. As the star ages along the main sequence, we expect the mean molecular weight μ to increase as hydrogen is fused into helium. Since the core temperature only increases modestly, the net effect is a decrease in c, concentrated near the core. This means that the D_0 parameter decreases with age in a predictable way. At the same time, the average large separation $\Delta\nu$, which is a function of the mean density ($\Delta\nu \propto \rho^{-1/2}$) decreases modestly, which suggests one way to track stellar evolution for low-mass stars on the main sequence is to plot the two parameters against one another, in what is known as an asteroseismic H–R diagram. This diagram was originated by Jørgen Christensen-Dalsgaard (JCD; Christensen-Dalsgaard 1988), and in his honor is frequently referred to as the *C-D diagram.* An example from Mazumdar (2005), based on stellar model calculations, is shown in Figure 6.3. The diagram shows the effects of mass and evolutionary state on the readily accessible asteroseismic parameters $\Delta\nu$ and $\delta\nu$, particularly below about $1.5M_{\odot}$. The large separation primarily reflects the mean density, while the small separation is most sensitive to conditions near the center, particularly the sound speed changes associated with the increase in mean molecular weight μ caused by stellar evolution.

The peaks themselves have nonzero width, and profiles that are Lorentzian in shape, which is to say

$$P(\nu) = \frac{1}{4}\frac{A^2}{4\pi^2(\nu - \nu_0)^2 + (1/\tau)^2} = \frac{1}{4}\frac{A^2}{4\pi^2(\nu - \nu_0)^2 + \eta^2} \tag{6.4}$$

where η is the inverse mode lifetime. This is what one would expect from the Fourier transform of a sinusoid like

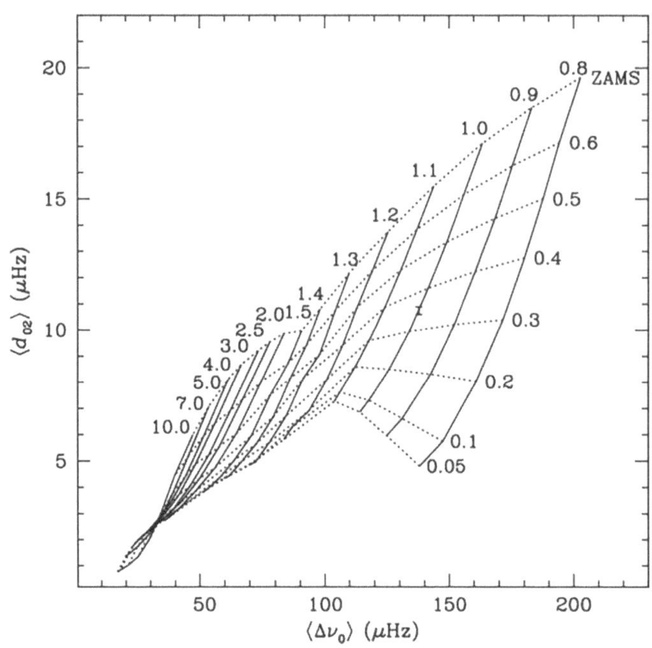

Figure 6.3. The asteroseismic H–R diagram, or C-D diagram, plotting the average small separation, $\delta\nu$, against the average large separation, $\Delta\nu$, for a grid of main sequence models between mass $0.8M_{\odot}$ and $10M_{\odot}$. The bold lines are evolutionary tracks for each mass as labeled, while the dotted lines connect models with the same central hydrogen abundance. Reprinted with permission from Mazumdar (2005).

$$y(t) = A\cos[2\pi(\nu t + \phi)], \tag{6.5}$$

but which persisted for only a finite period in time, τ. Note that this finite time could be the *lifetime* of the oscillation mode, **or** it could simply be the length of the time series itself. The characteristic width of this Lorentzian profile in the frequency domain is just

$$\Gamma = \eta/\pi \tag{6.6}$$

But how do we actually measure these parameters?

6.1.2 Peak-Bagging

Some hikers engage in the activity of "peak-bagging," where they attempt to summit as many mountains as possible in a given region. The term has been adopted by asteroseismologists as a colloquial expression for the process of finding the maximum number of oscillation frequencies ("peaks") in the power spectrum of a star. As with its mountain-climbing cousin, seismic peak-bagging can be competitive in the sense that algorithmic or instrumental improvements can raise the number of significant peaks detected, which of course in turn improves the quality of inferences that can be drawn from the time series.

The fundamental tool for determining the frequency spectrum of an oscillating star is the power (or amplitude) spectrum and its variants. From that spectrum, we measure the frequencies themselves, which we use in turn to derive the physical parameters of the star. Initially, frequency identification in the spectra was performed more or less "by eye," with frequencies measured using simple single Lorentzian fits to the handful of detectable lines, or even simply read off from the graph. The arrival of long-duration, high signal-to-noise time series from space has resulted in the advent of power spectra containing large numbers of peaks that are no longer conducive to such boutique measurement and call for more automated and standardized techniques. For massive stars with coherent long-lived oscillations, fitting a series of sinusoids to the data can be sufficient, but for solar-like stars such an approach only scratches the surface. In addition, particularly for solar-like stars, the individual mode peaks lie on a background, due primarily to stellar granulation and supergranulation, that rises at low frequencies, complicating both the fitting process and determination of mode amplitudes and positions. Furthermore, as we have seen, there are additional contributors to the power spectrum background, including stellar activity, various instrumental noise contributions, and of course photon noise. Beyond simply measuring its location and amplitude, how do we know that a detected peak is significant, and that it represents a frequency of interest rather than—for example—some instrumental contribution (Figure 6.4)?

The earliest approaches for asteroseismology, exemplified by software tools such as Period98 and later Period04 (Sperl 1998; Lenz & Breger 2014), focused on oscillations in massive stars and made use of a combination of the discrete Fourier transform to determine the approximate frequencies of peaks, followed by least-squares fitting to multiple frequencies to determine amplitudes, phases, and formal

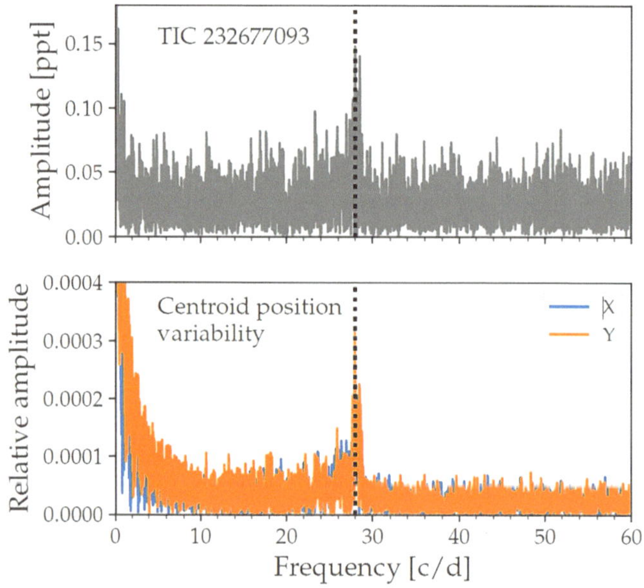

Figure 6.4. Bottom: Jitter induced by a quasi-periodic heating element switching on and off on the TESS spacecraft, shown as regular variation in centroid position. Top: The jitter, by moving the stellar image with respect to the photometric aperture, induces periodic behavior in the apparent flux of the target.

uncertainties. With this approach, the background is assumed to be white (constant in power as a function of frequency) and significance is taken as being determined by the signal-to-noise ratios of the individual peaks relative to the local background (typically SNR = 4 is adopted as the cutoff). While this can be a reasonable approach in cases where the oscillation frequencies are solidly in the high-frequency, white noise, portion of the spectrum, at shorter frequencies where the pink and red components of the noise (instrumental and stellar) are contributing rising proportions of the total noise, it becomes more and more likely that noise peaks are present that are higher than any reasonably-chosen SNR threshold. In addition, since oscillation modes in solar-like stars have mode lifetimes which are short compared to the time series length, they appear broadened in the power spectrum, with their power distributed over several frequency resolution bins. This in turn lowers the peak height and will cause real frequency peaks to be missed in the analysis (Figure 6.5).

The primary contributors to the background level for space-based data are stellar contributions such as granulation and supergranulation. In both cases, the contribution can be parameterized in terms of timescale (τ_{gran}) and amplitude (σ_{gran}). Most background models used draw on the approach of Harvey (1985), who approximated the autocovariance of granulation evolution with an exponential decay function making use of the two parameters. This leads to a Lorentz profile power spectrum of the form

$$P(\nu) = \frac{4\sigma^2\tau_{gran}}{1 + (2\pi\nu\tau_{gran})^{\alpha}} \tag{6.7}$$

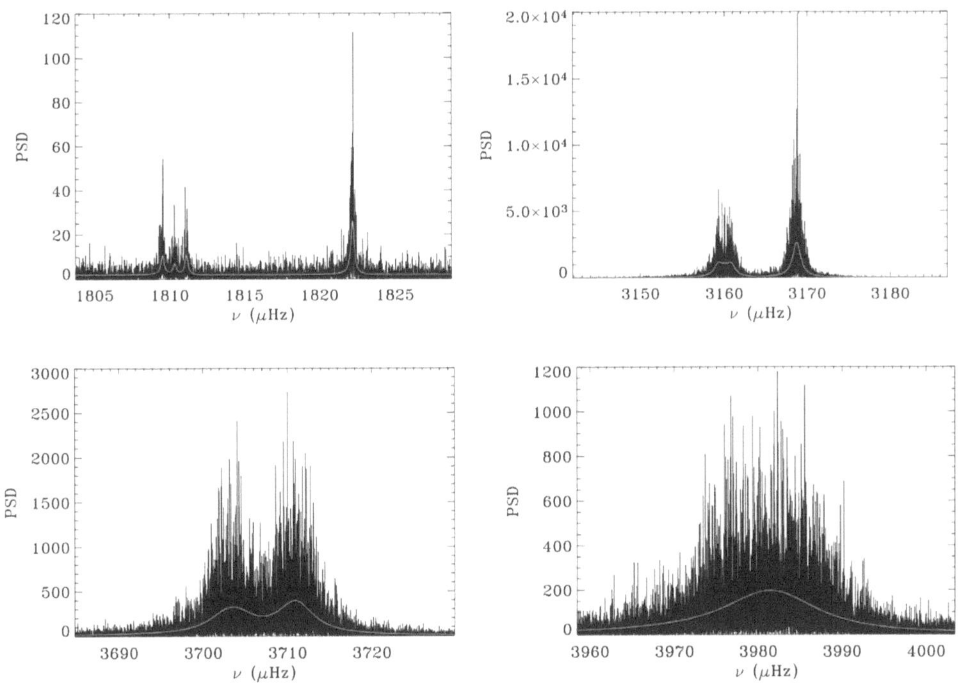

Figure 6.5. Fits to $l, m = 2, 0$ and $l, m = 3, 1$ frequency peaks from the solar oscillation spectrum, derived from the GOLF instrument. The longer lifetimes of the 2, 0 modes are reflected in their narrower line widths. Reprinted with permission from García & Ballot (2019).

In Harvey's initial formulation $\alpha = 2$, but later investigators have found that allowing that parameter to vary improves the representation. It also changes the normalization factor, so

$$P(\nu) = \frac{4\xi\sigma^2\tau_{\text{gran}}}{1 + (2\pi\nu\tau_{\text{gran}})^\alpha} \tag{6.8}$$

where $\xi = 2\alpha\,\sin(\pi/\alpha)$ (Karoff et al. 2013). Different components are used for granulation and supergranulation, as seen in Figure 6.6, where a good fit requires three Harvey profiles, representing three different granulation/supergranulation scales, together with the photon noise. Granulation parameters are related to physical conditions. Huber et al. (2009) suggest that the granulation timescale τ_{gran} is proportional to the pressure scale height and inversely proportional to the sound speed, so the timescale can be expressed as (Huber et al. 2009)

$$\tau_{\text{gran}} \propto \frac{H}{c} \sim \frac{L}{T_{\text{e}}^{3.5} M} \sim \nu_{\max}^{-1} \tag{6.9}$$

A second granulation parameter, the photometric amplitude, can be expressed as

$$\sigma_{\text{gran}} \propto \frac{c}{\sqrt{n}} \tag{6.10}$$

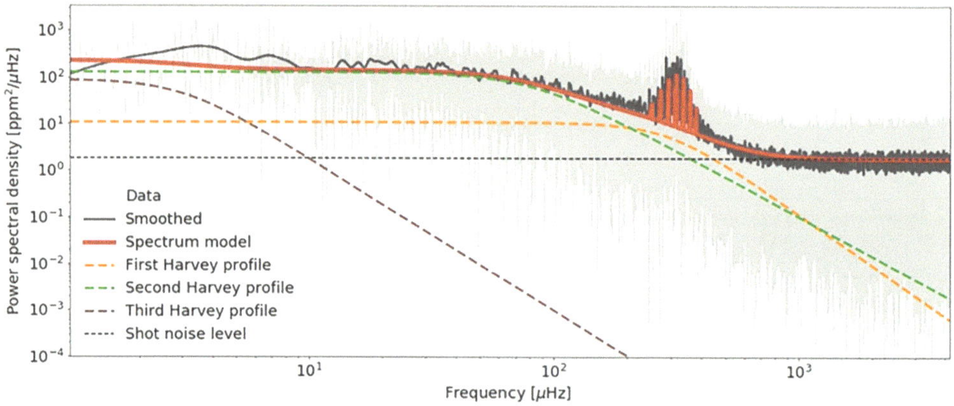

Figure 6.6. Power spectrum of the red giant ε Reticuli. The raw PSD is shown in light gray and the smoothed in dark gray. The various components of the background model are shown with dashed lines, while the combined spectrum model, including the mode components, is shown in red. Reprinted with permission from Nielsen et al. (2023).

where n is the number of surface granules, which we anticipate to be

$$n \sim \left(\frac{R}{H}\right)^2 \tag{6.11}$$

Together these imply

$$\sigma_{\rm gran} \sim \frac{L^{0.5}}{MT^{0.5}} \tag{6.12}$$

Ren and collaborators (Ren & Jiang 2020; Ren et al. 2024) determined that a better fit to the data for red supergiants was $\sigma_{\rm gran} \propto L^{0.75}/T^{2.5}$, but showed indications that the fit extended successfully all the way to solar-like stars. Though there is clearly more work to be done, it is clear that the granulation fits incorporate useful global stellar parameters.

6.2 Average Seismic Parameters

Global seismic parameters can be accessible even when when the individual mode peaks have low signal to noise, and even when this is not the case they are much easier and faster to measure and analyze, particularly for the larger data sets now available from space missions. The two most commonly used such parameters are $\nu_{\rm max}$, the frequency of maximum oscillation power, and $\overline{\Delta\nu}$, the average large frequency separation.[2]

[2] Confusingly, this is frequently written as simply $\Delta\nu$, and we will perpetuate this confusion going forwards.

6.2.1 Frequency of Maximum Power

For solar-like oscillators, we observe that the power excess due to oscillations in the power or amplitude spectrum appears to be roughly modulated by a Gaussian curve, as shown in Figure 2.6.

The localization of oscillation power in frequency space is imposed by the interaction between the oscillation excitation and damping mechanisms (see, e.g., Hekker & Christensen-Dalsgaard 2017 and references therein for a more detailed discussion). Nearly a century ago, Lamb (1932) pointed out that for wave propagation in an isothermal vertically stratified atmosphere, the characteristic timescale is the acoustic cutoff frequency, or

$$\nu_{\rm ac} = \frac{c}{4\pi H} \tag{6.13}$$

where H is the atmospheric pressure scale height, or

$$H = \frac{kT}{\mu m_{\rm H} g} \tag{6.14}$$

Brown et al. (1991) argued that the frequency of maximum power should be approximately of this scale, and this appears empirically to be the case. Since

$$\nu_{\rm ac} \sim \nu_{\rm max} \propto \frac{g}{\sqrt{T_{\rm eff}}} \propto \frac{M}{R^2\sqrt{T_{\rm eff}}} \tag{6.15}$$

Observationally, $\nu_{\rm max}$ represents the frequency at which the signal-to-noise ratio of the power spectrum is a maximum, at least in the frequency region immediately surrounding the Gaussian-shaped envelope describing the peaks for solar-like oscillators. For solar-like stars on the main sequence, this peak corresponds to a radial order in the neighborhood of $n = 20$, though both the value of $\nu_{\rm max}$ and the corresponding radial order decrease as the star evolves and its density drops, reaching radial order values of only a couple by the time it reaches the tip of the red giant branch.

The actual method for measuring ν_{max} varies by practitioner. Most begin by removing the granulation and/or supergranulation signals by fitting and subtracting Harvey profiles. Perhaps the simplest approach then is to use the peak of the excess power in the smoothed power spectrum (Huber et al. 2009), though a choice must then be made in how to smooth. The optimal width of the smoothing function should be proportional to the large frequency separation $\Delta\nu$, but the exact dependence adopted varies. Typical values are $1 - 2\Delta\nu$, though some algorithms are more complex; for example the commonly-used SYD pipeline (Huber et al. 2009; Chontos et al. 2021) uses

$$\Delta\nu \times \max\left[1, 4\left(\frac{\nu_{\rm max}}{\nu_{\rm max,\odot}}\right)^{0.2}\right] \tag{6.16}$$

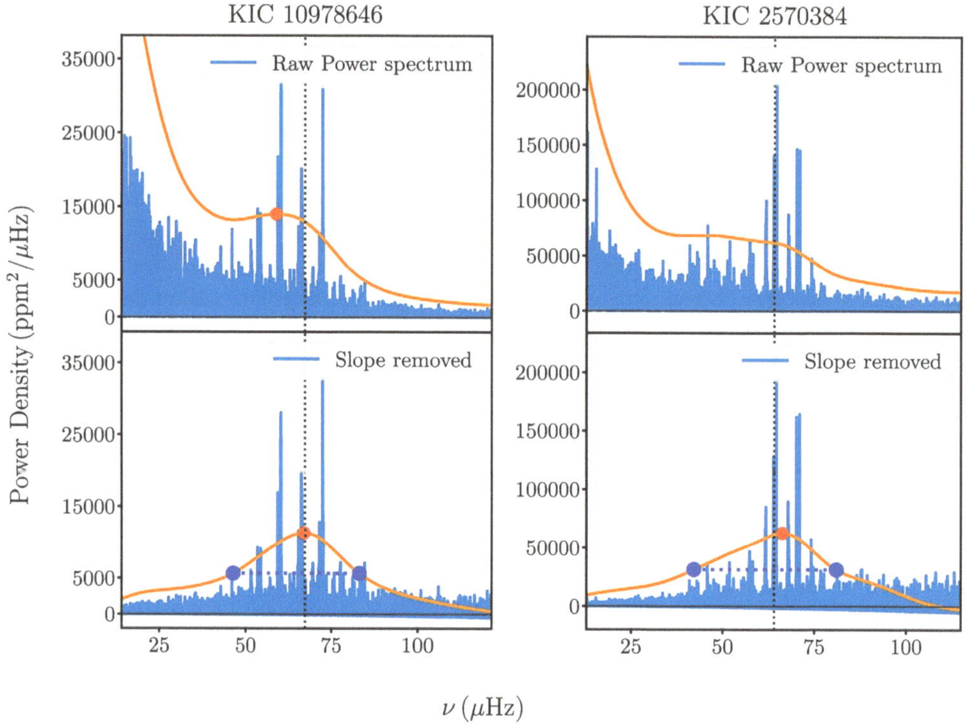

Figure 6.7. An example of measuring ν_{max}. Here the orange curves show smoothed power spectra (scaled in the vertical direction for better visibility) and the red dots show the ν_{max} values. Upper panels are without removing the background slope, while lower panels show the final result after removing background fits by division. The blue dashed lines connect the two half-power points, and the vertical black dashed line shows ν_{max} previously reported in the literature. Reprinted with permission from Sreenivas et al. (2024).

Another alternative is to use to fit a Gaussian to the oscillation power excess and then adopt the centroid of the fit (Kallinger et al. 2010) as ν_{max} and estimate the uncertainty in that location from the width of the Gaussian, such that

$$\sigma_{\nu_{max}} \sim \frac{\text{FWHM}}{\text{A}}, \tag{6.17}$$

where A is the peak amplitude. More sophisticated algorithms include using the first moment of the area under the smoothed power spectrum (Hekker et al. 2010) or wavelet transforms. Alternate methods deal with the granulation signature by dividing the power spectrum by a functional form and then smoothing (Sreenivas et al. 2024; see Figure 6.7), making use of a collapsed 2D autocorrelation function (Kiefer 2018), or by applying machine learning analysis (Hon et al. 2018). Bell et al. (2019) suggest using what they term the coefficient of variation (CV), the ratio of the standard deviation to the mean of the power spectrum; regions of the power spectrum containing oscillations are those where CV $>$ 1. Typical internal uncertainties are $1 - 2\%$ (Sreenivas et al. 2024; Viani et al. 2019), but systematic offsets between different methods are of the same order of magnitude, arising in many cases from the differing methods of accounting for the underlying background power.

6.2.2 Mean Large Separation

The mean large frequency separation between modes of consecutive radial order n and the same spherical degree l is directly related to the sound travel time across the star. It therefore samples the entire stellar interior, and is sensitive to the mean stellar density (Ulrich 1986).

We can understand this by looking at the dynamical timescale for the star, which is

$$t_{\rm ff} = \left(\frac{2R^3}{GM}\right)^{1/2} \tag{6.18}$$

To within some factor of order unity, we expect this to express the fundamental oscillation frequency for the star, and higher-order oscillation modes to lie above this frequency. Of course the precise frequencies depend, as we will see later (and in Chapter 8), on details of the internal stellar structure, but the implication of the simple relation that

$$\nu_{\rm max} \sim \left(\frac{M}{R^3}\right)^{1/2} \sim \bar{\rho}^{1/2} \tag{6.19}$$

is that the spacing between successive oscillation modes scales similarly, so the mean large separation should behave as

$$\Delta\nu \sim (\Delta\nu)_{\odot}\left[\frac{M/M_{\odot}}{(R/R_{\odot})^3}\right]^{1/2} \tag{6.20}$$

In practical terms, measuring the large frequency separation entails searching for periodicities in the power spectrum itself, focusing on the region immediately surrounding $\nu_{\rm max}$. In the high signal-to-noise case, this can be done by fitting and measuring individual frequencies, and then either fitting the frequencies versus radial order or examining pair-wise differences. However, more commonly the SNR is low, which favors approaches such as autocorrelation of the power spectrum (Roxburgh & Vorontsov 2006; Huber et al. 2009; Mosser & Appourchaux 2009) or calculating the power spectrum of the power spectrum (PS $\otimes$ PS) (Hekker et al. 2010; Mathur et al. 2010). In either case, the frequency range to be examined must be limited to avoid dilution of the signal by the portions of the power spectrum that don't contain oscillation signals. This is generally derived from the $\nu_{\rm max}$ fit and its uncertainty $\sigma_{\nu_{\rm max}}$. An excellent first guess at the value of the large separation can be derived using the power-law relationship between $\Delta\nu$ and $\nu_{\rm max}$ as discussed in Section 6.5 and shown in Figure 6.12. Quoted uncertainties in $\Delta\nu$ are typically of order 0.1μHz or less.

Table 6.1 shows an overview of a number of popular codes for deriving global asteroseismic parameters.

Table 6.1. Methods Used by Popular Codes to Derive Global Asteroseismic Parameters $\nu_{\rm max}$ and $\Delta\nu$

Name	$\nu_{\rm max}$ Method	Smoothing?	$\Delta\nu$ Method	ν Interval	Background	Notes
A2Z (Mathur et al. 2010)	Gaussian centroid	$1-2\Delta\nu$ box	$PS \otimes PS$	Excess PS power	White noise (WN) + power law + 1 Harvey	Gap-filling
BAM (Zinn et al. 2019)	Gaussian centroid	$2\Delta\nu$ Gaussian	Autocorrelation or folded PS	Prior based on $\nu_{\rm max}$	WN + 2 Harvey, apodized	Bayesian; light curve is smoothed and gap-filled
BHM (Stello et al. 2017)	Gaussian centroid	$2\Delta\nu$ box	$PS \otimes PS$	FWHM of excess peak	WN	Background removal by division of smoothed PS
CAN (Kallinger et al. 2010)	Gaussian centroid	None	Fit to PS (8 frequencies)	$\nu_{\rm max} \pm 2\sigma_G$	WN + 3 Harvey	
COR (Mosser & Appourchaux 2009)	Gaussian centroid	$3\Delta\nu$ Gaussian	Time series autocorrelation	FWHM of excess peak	Power law	
OCT (Hekker et al. 2010)	Gaussian centroid or first moment	$2\Delta\nu$ box	$PS \otimes PS$	FWHM of excess peak	Linear	
SYD (Huber et al. 2009)	Peak smoothed PS	$2\Delta\nu$ Gaussian	Autocorrelation PS	$\nu_{\rm max} \pm 10\Delta\nu$	WN + modified Harvey	Number of Harvey components adjustable

6.2.3 Scaling Relations

The value of measurements of $\nu_{\max}$ and $\Delta\nu$ lies in their relation to fundamental stellar properties. If we treat the star as a blackbody radiator, then

$$L \sim R^2 T^4 \tag{6.21}$$

and substituting this into the expression above for $\nu_{\max}$ gives

$$\nu_{\max} \sim \frac{MT^{3.5}}{L} \tag{6.22}$$

This simple relation allows us to estimate the stellar mass based on $T_{\rm eff}$ and luminosity L, both of which are relatively straightforward to obtain using spectroscopy and distance measurements from *Gaia*. Reorganizing and scaling based on solar parameters gives straightforward expressions of the form

$$\nu_{\max} \sim \left(\frac{M}{M_\odot}\right)\left(\frac{R}{R_\odot}\right)^2\left(\frac{T_{\rm eff}}{T_{\rm eff,\odot}}\right)^{0.5} \tag{6.23}$$

$$\Delta\nu \sim \left(\frac{M}{M_\odot}\right)^{0.5}\left(\frac{R}{R_\odot}\right)^{-1.5} \tag{6.24}$$

More usually, we solve for the stellar physical parameters, to get

$$\frac{M}{M_\odot} = \left(\frac{\nu_{\max}}{\nu_{\max,\odot}}\right)^3\left(\frac{\Delta\nu}{\Delta\nu_\odot}\right)^{-4}\left(\frac{T_{\rm e}}{T_{\rm e,\odot}}\right)^{1.5} \tag{6.25}$$

$$\frac{R}{R_\odot} = \left(\frac{\nu_{\max}}{\nu_{\max,\odot}}\right)\left(\frac{\Delta\nu}{\Delta\nu_\odot}\right)^{-2}\left(\frac{T_{\rm e}}{T_{\rm e,\odot}}\right)^{0.5} \tag{6.26}$$

6.3 Echelle Diagrams

Knowing the average large separation allows us to make and use a simple analytical tool, the so-called *echelle diagram*. To construct the diagram, we simply cut the power spectrum into pieces of length $\Delta\nu$ and create a two-dimensional visualization by stacking the pieces on top of one another. Figure 6.8 shows examples for three stars (Metcalfe et al. 2014). The diagram visibly accentuates the structure in the power spectrum, and makes departures from strictly periodic peak separation obvious as departures from perfect alignment. Each marked point represents a single oscillation mode, and each vertical ridge shows the oscillations with different radial order n but the same degree l.

The basic structure of the diagram is defined by the asymptotic relations, so we can write

$$\nu_{nl} = \Delta\nu\left(n + \frac{l}{2} + \varepsilon\right) - \delta\nu_{0l} \tag{6.27}$$

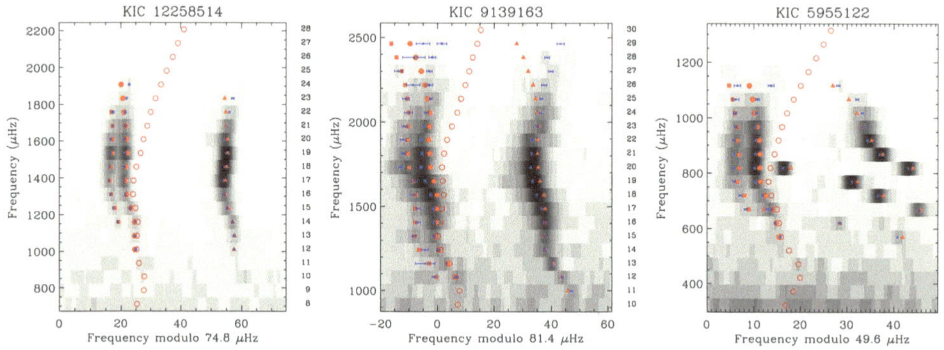

Figure 6.8. Echelle diagrams for three different types of stars. On the left is a simple lower main sequence star, at center an F star, and on the right a star oscillating in mixed modes. Each plot shows a smoothed grayscale representation of the power spectrum overplotted with frequencies and errors from multiple models, including (open red circles) models uncorrected for the surface effect and from the surface-corrected AMP model (solid red points). Reprinted with permission from Metcalfe et al. (2014).

The terms have physical significance, through their relationship to the sound speed and its gradients. The large separation itself is proportional to the square root of the mean stellar density, while in main sequence stars the small separation is sensitive to sound speed gradients in the core, while $\delta\nu_{0l} \propto l(l+1)$. ε reflects surface conditions; more specifically, the upper turning point of each mode. ε can in theory be read off directly from the echelle diagram, but in practice if the radial order n is unknown Equation (6.27) indicates that the choices of n and ε are degenerate, and only in case of the Sun do we usually know the values of n with complete certainty.

Clearly, if frequency spacings were strictly regular, so that the large and small separations both represented exact rather than approximate descriptions of the frequency locations, the proper choice of $\Delta\nu$ in the plot would result in precisely vertical ridges. In reality, ridges are always curved to lesser or greater degree, and the shapes as well as the locations of the ridges encode information about the stellar structure, making the echelle diagram a useful diagnostic tool. The diagram shown in the left hand panel of Figure 6.8 shows the typical S-shaped curve described by the modes, which is caused by both variations in $\Delta\nu$ and the surface parameter ε as functions of frequency. The pattern of ridges allows identification of the $l = 1$–4 modes, though hotter main-sequence stars tend to have larger intrinsic line widths, blurring the individual modes, broadening the ridges, and potentially complicating mode identification when the ridges begin to overlap (Figure 6.8, center panel).

The diagram also shows changes with stellar evolution. Most obviously, as stars evolve on the main sequence, their surface gravity decreases modestly, leading both $\Delta\nu$ and ν_{max} to move to lower frequencies; on the echelle diagram this shows as both a decrease in the $\Delta\nu$ needed to construct the diagram and a decrease in the spacing between ridges. Note that the diagram also shows the small frequency separations, both between $l = 0$ and $l = 2$ modes with the same n, $\delta\nu_{02}$ and between $l = 0$ and $l = 3$ modes with the same n, $\delta\nu_{03}$. These small separations are functions of the sound speed gradient in the stellar interior; for solar-like stars on the main sequence, this provides an age diagnostic as they respond to the composition gradient in the core caused by the changes in mean molecular weight μ caused by nuclear fusion over time.

6.4 Massive Stars

In low-mass stars, pressure modes occur in the convective outer envelope and have relatively large amplitudes in near-surface layers, while gravity modes are confined to the radiative core and are evanescent in the envelope, making them extremely difficult to detect from outside the star. However, in more massive stars with radiative outer layers, gravity modes exist in large amplitudes in the radiative cavity and are readily detectable. They are therefore the dominant asteroseismic tool for the study of massive stars.

Gravity modes, relying on gravity as their restoring force, have frequencies lower than the Brunt–Väisälä frequency N, where

$$N^2 = g\left(\frac{1}{\gamma P}\frac{dp}{dr} - \frac{1}{\rho}\frac{d\rho}{dr}\right) \tag{6.28}$$

As in the case of p-modes, in the asymptotic regime, high radial order gravity modes are approximately equally spaced, only here in period rather than frequency, with a characteristic period spacing Π_0

$$\Pi_0 = 2\pi^2\left(\int_{r_1}^{r_2} N(r)\frac{dr}{r}\right)^{-1} \tag{6.29}$$

where the inner and outer radii r_i define the limits of the pulsation cavity. The asymptotic period Π_0 itself is also related to the individually measured gravity mode periods through

$$P_{nl} = \frac{\Pi_0}{\sqrt{l(l+1)}}(|n| + \alpha) \tag{6.30}$$

where α is an l-independent phase term.

Rotation in these stars is visible in the structure of the peaks themselves. Mode frequencies are affected by rotation through removal of the frequency degeneracy in the quantum number m, so for relatively slow rigid-body rotation,

$$\omega_{nlm} = \omega_{nl} + m(1 - C_{nl})\Omega \tag{6.31}$$

where Ω is the stellar angular rotation rate and $0 \leqslant C_{nl} \leqslant 1$ is the *Ledoux constant*; its exact value depends on the mode geometry nlm and the internal structure of the star, but for $l = 1$, it's typically around 0.5. There are $2l + 1$ values of m for each l, so peaks are effectively split into multiplets based on l, which aids in mode identification. Figure 6.9 shows an example. Such observations are possible for solar-like oscillators as well, but are greatly complicated by the generally slower

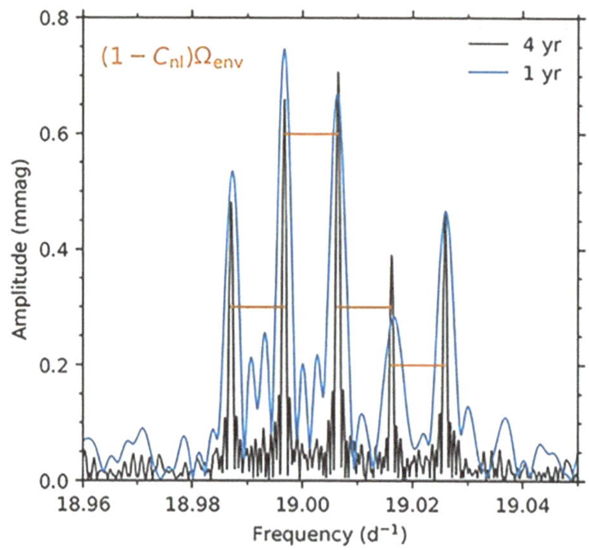

Figure 6.9. Rotational splitting of an $l = 2$ (quadrupole) p-mode into a quintuplet. The two overlying curves indicate the spectrum after 1 and 4 years of Kepler data on the star KIC 1114512 (Kurtz et al. 2014). Reprinted with permission from Bowman (2020a).

rotation rates of those stars (leading to smaller rotational splittings) and the shorter mode lifetimes of the stochastically-excited modes in cool stars (leading to broader line widths). However, some successes have been reported there as well (Nielsen et al. 2020).

The presence of magnetic fields affects the mode frequencies as well, though in this case a weak magnetic field shifts them to higher frequencies, so that

$$\omega_{nlm} = \omega_{nl} + \frac{m^2 B^2}{\omega_{nl}^3} R_{nl} \tag{6.32}$$

Here R_{nl} is another structure constant of order unity. Note that the dependence on ω_{nl}^{-3} means that, if rotation is also splitting a frequency peak, the resultant multiplet will be asymmetrical if magnetic fields are present as well.

The regularity of period spacings allows construction of a *period spacing diagram*, showing the period differences of modes of the same angular degree l and azimuthal order m but consecutive radial order n (Figure 6.10). The alternating pattern visible between consecutive periods is due to the sharp mean molecular weight gradient near the core, though it can be washed out in stars with rapid rotation and/or strong diffusion effects.

The period spacing diagram calls out departures from the regular pattern. Departures are caused by two main effects. First, the pattern shows a regular overall slope, which is due to rotation; more rapid rotation *ceteris paribus* causes a

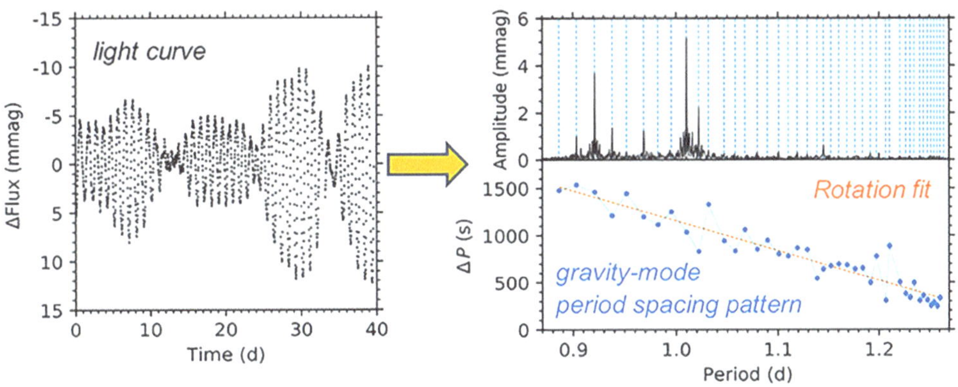

Figure 6.10. Left: Kepler light curve of the Slowly Pulsating B (SPB) star KIC 3459297 pulsating in low-frequency g modes. Beating between the two highest-amplitude modes is readily apparent. Right: Amplitude spectrum (top) and period-spacing pattern (bottom) for the light curve shown on the left. Prograde g-mode pulsations are visible, and imply a rotation rate of $f_{\rm rot} = 0.63 \pm 0.04 \mathrm{d}^{-1}$. Reprinted with permission from Bowman (2020a).

steeper slope. This is because the effective $\mathbf{L}^2$ increases (decreases) with rotation for prograde (retrograde) modes due to the effects of the Coriolis force, leading to the period spacing to increase (decrease) with rotation for prograde (retrograde) modes. If the regular slope is removed, somewhat irregular bumps remain in the structure; these are due to spherical shells with chemical composition gradients, which manifest themselves as changes to the local sound speed. Unfortunately, the bumps do not uniquely determine composition gradients, which also depend on mass, age, and internal mixing.

Bowman et al. (2019) and Bowman (2020b) reported the ubiquitous presence of low-frequency stochastic variability in the most massive stars; the variability appears as a low-frequency granulation-like noise excess in stars, which of course lacks surface granulation. Bowman et al. (2019) and Bowman (2020b) ascribe the source of this variability, which correlates with macroturbulent line broadening observed in spectroscopy, to internal gravity waves excited by core convection. The morphology of the noise signature in the frequency domain probes bulk stellar properties such as mass and radius, and it provides a window into greater understanding of mixing and angular momentum transport at the top of the main sequence.

δ Scuti stars are examples of more massive stars that lie on the instability strip, and thus show oscillations of relatively large amplitude. However, understanding of these stars has been hampered by the fact that, as with other massive pulsators, most modes near some $\nu_{\rm max}$ are not excited (as with solar-like oscillators), but instead an apparently random subset of possible modes shows large amplitudes.[3] This lack of

[3] Of course, the apparently random nature of this sample really only reflects our lack of detailed knowledge of the excitation and damping mechanisms, and perhaps the internal structure, of these stars.

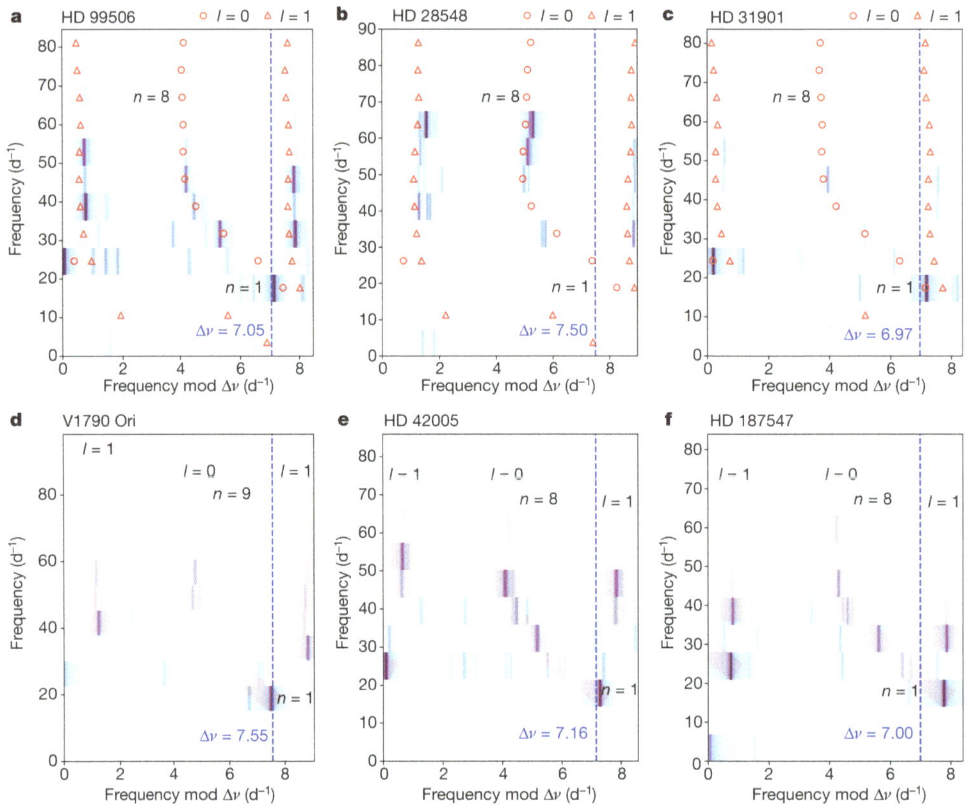

Figure 6.11. Pulsation spectra of a sample of δ Scuti stars, shown as echelle diagrams, with a repeated overlap region added on the right for clarity. Data are from the TESS satellite. In the top row, red symbols show frequencies calculated from non-rotating models, while in the bottom row the red stripes mark overtone sequences of $l = 0$, 1 modes. Reprinted with permission from Bedding et al. (2020).

regularity can make it difficult to assign quantum numbers *nl* to each peak, which makes modeling the spectrum difficult. However, some δ Scuti stars do show relatively regular patterns, which can be seen in echelle diagrams (Figure 6.11), and can lead to mode identification as well as the estimation of the mean stellar density.

6.5 Evolved Stars

Scaling relations continue to work well for red giants, as seen in Figure 2.7. The p-mode spectrum continues to show a Gaussian distribution of peaks, which leads to a measurable $\nu_{\max}$ and large frequency separation $\Delta\nu$. In fact, the scaling laws would predict a relationship between those two parameters, and observationally this is confirmed, from the coolest main sequence stars in which oscillations have been detected (early K dwarfs) up through the red giant branch (Figure 6.12). A more detailed look at the actual frequencies seen in evolved stars shows some detailed changes, however.

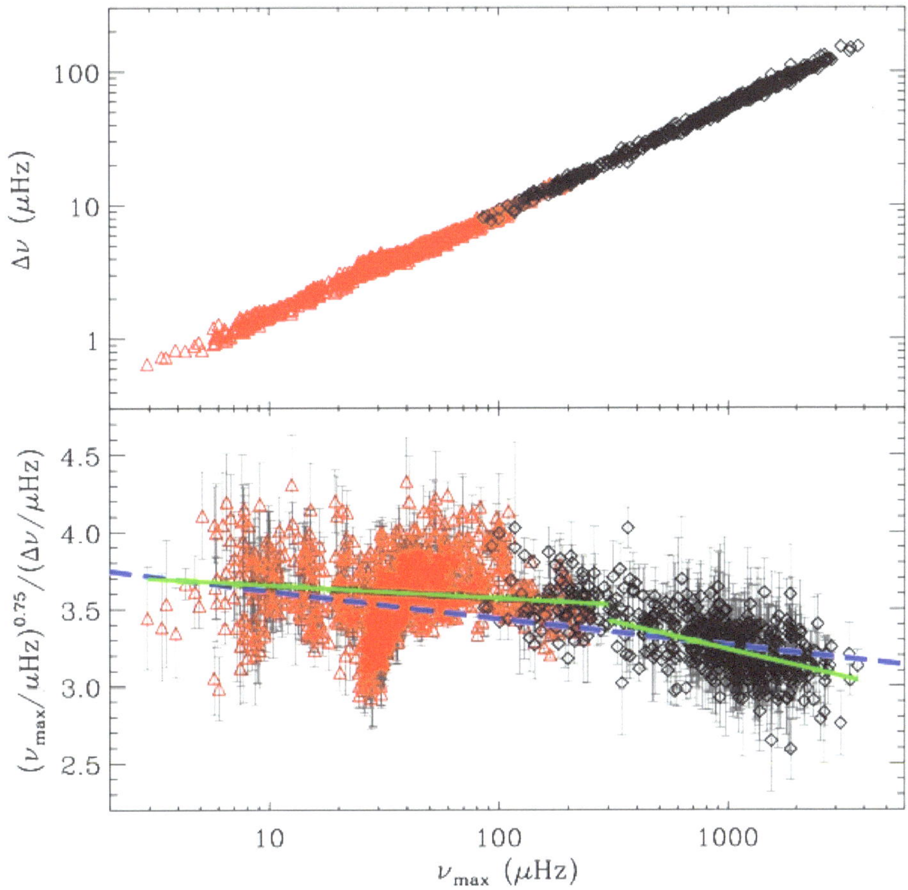

Figure 6.12. Upper panel: $\Delta\nu$ versus $\nu_{\rm max}$ for over 1700 stars observed by the Kepler Mission, spanning the region from the main sequence to the red clump. Red triangles indicate long-cadence observations and black diamonds short cadence. Lower panel: A power-law fit ($\Delta\nu \propto \nu_{\rm max}^{0.75}$) has been removed, showing that the relation steepens for $\nu_{\rm max} > 300\ \mu{\rm Hz}$. The green lines in the lower panel show a broken power-law fit, while the blue dashed line shows the relationship from Stello et al. (2009). Reprinted with permission from Huber et al. (2011).

As noted above, the echelle diagram can illustrate the evolution of solar-like stars off the main sequence. As stars begin to move off the main sequence and become subgiants, modes appear of mixed *p* and *g* character ("mixed modes"), which couple the g-mode cavity in the core with the p-mode oscillation cavity in the envelope. This produces the phenomenon of *avoided crossings* (Aizenman et al. 1977), analogous to what is observed with coupled mechanical oscillators. Mechanically, with two independent uncoupled oscillators, the physical parameters of one member of the pair can be tuned to exactly match the other. Once the oscillators are allowed to interact, this is no longer true, and varying the parameters of one (or both) oscillators cannot lead to a "crossing" of the resonant frequencies of the two cavities. The strength of the avoidance is a function of the strength of the coupling of the two oscillators; in the stellar case, the coupling between the inner and outer resonant cavities through the evanescent zone that separates them.

An example of such an avoided crossing is visible in the right-hand panel of Figure 6.8, which illustrates the echelle diagram for a subgiant. The departure from regularity is obvious, and shows us that the resonant cavities lie close to one another, which indicates that the star has begun to evolve off the main sequence. The way in which the regular pattern in the echelle diagram is disrupted is a diagnostic for the otherwise-inaccessible core region of the star, as the mixed modes imprint the period-spacings from the g-modes into the p-mode pattern we **can** observe, as shown earlier in Figure 2.11.

6.6 Forward Modeling

Measurement of parameters such as $\nu_{\max}$ and $\Delta\nu$, along with C-D diagrams, echelle diagrams, period spacing diagrams, and measurements of rotational splittings can provide great insight into the interior structure and behavior of oscillating stars, but are only a start. For detailed understanding, we need to calculate interiors models that display the same oscillatory behavior as the actual stars we observe.

Broadly speaking, there are two approaches. First, we can attempt to *invert* the frequencies to infer the star that produced them. This has been successfully done for the Sun and a (very small) handful of other stars, and we will discuss it below in Section 8.1, but it poses significant difficulties. Alternatively, we can attempt to *forward model*, by calculating oscillation frequencies for a selection of models and attempting to vary the input models to match observations. The complication here is that there are numerous variables to worry about: mass, radius, age, luminosity, composition and composition gradient, mixing length parameter, rotation, diffusion, and others, while the uncertainties on the asteroseismic line frequencies are typically quite small. This means that the most common approach, *grid modeling*, where a wide range of models spanning all reasonable physical parameters is calculated, and generally interpolated to reproduce observations.

Such an approach is computationally extremely expensive, as it can easily lead to the calculation of hundreds of thousands of models. The use of additional constraints on stellar parameters, e.g., from spectroscopy, or cluster membership or space velocities, or binarity, can collapse the parameter space sufficiently to make the problem more tractable. For low-mass stars and giants, the scaling relations can bound at least some of the parameters of interest, though these are not any help for higher-mass stars. Faster modern computers help, too. Table 6.2 shows a selection of available codes for calculating oscillation properties based on an input stellar model. Some are intended to be for general application, while others are optimized for particular types of stars, or for use with particular modeling software.

Table 6.2. A Selection of Commonly-Used Pulsation Codes

Name	Types	Radial/ Non	Notes
ADIPLS (Christensen-Dalsgaard 2008)	A	R/N	Aarhus adiabatic oscillation package
FILOU (Suran 2008)	A	R/N	Optimized for classical pulsators
GraCo (Moya & Garrido 2008)	A/N	R/N	Granada oscillation code
GYRE (Townsend & Teitler 2013)	A/N	R/N	Commonly used with MESA
JIG (Guenther 1994)	A/N	R/N	Optimized for use with YREC
LOSC (Scuflaire et al. 2008)	A	R/N	Liege Oscillations Code
LNAWENR (Suran 2008)	A/N	R/N	Can determine color variations
MESA-RSP (Paxton et al. 2019; Smolec & Moskalik 2008)	A/N	R	Built into MESA, classical pulsators
NOSC (Provost 2008)	A	R/N	Nice Oscillations Code
OSCROX (Roxburgh 2008)	A	R/N	low-degree modes
PULSE (Brassard & Charpinet 2008)	A/N	R/N	Originally developed for white dwarfs
POSC (Monteiro 2008)	A	R/N	Porto Oscillation Code; PMS to subgiant

References

Aizenman, M., Smeyers, P., & Weigert, A. 1977, A&A, 58, 41

Bedding, T. R., & Kjeldsen, H. 2003, PASA, 20, 203

Bedding, T. R., Murphy, S. J., Hey, D. R., et al. 2020, Natur, 581, 147

Bell, K. J., Hekker, S., & Kuszlewicz, J. S. 2019, MNRAS, 482, 616

Bowman, D. M. 2020a, FrASS, 7, 70

Bowman, D. M. 2020b, FrASS, 7, 70

Bowman, D. M., Aerts, C., Johnston, C., et al. 2019, A&A, 621, A135

Brassard, P., & Charpinet, S. 2008, ApS&S, 316, 107

Brown, T. M., Gilliland, R. L., Noyes, R. W., & Ramsey Lawrence, W. 1991, ApJ, 368, 599

Chontos, A., Huber, D., Sayeed, M., & Yamsiri, P. 2021, pySYD: Measuring Global Asteroseismic Parameters, Astrophysics Source Code Library, ascl:2111.017

Christensen-Dalsgaard, J., & Frandsen, S. 1988, IAU Symp., Vol. 123, Advances in Helio- and Asteroseismology ed. J. Christensen-Dalsgaard, & S. Frandsen (Dordrecht: Reidel Publishing) 295

Christensen-Dalsgaard, J. 2008, ApS&S, 316, 13

García, R. A., & Ballot, J. 2019, LRSP, 16, 4

Guenther, D. B. 1994, ApJ, 422, 400

Harvey, J. 1985, ESA Special Publication, Vol. 235, Future Missions in Solar, Heliospheric & Space Plasma Physics ed. E. Rolfe, & B. Rolfe (Paris: ESA) 199

Hekker, S., Broomhall, A. M., Chaplin, W. J., et al. 2010, MNRAS, 402, 2049

Hekker, S., & Christensen-Dalsgaard, J. 2017, A&ARv, 25, 1

Hon, M., Stello, D., & Zinn, J. C. 2018, ApJ, 859, 64

Huber, D., Stello, D., Bedding, T. R., et al. 2009, CoAst, 160, 74
Huber, D., Bedding, T. R., Stello, D., et al. 2011, ApJ, 743, 143
Kallinger, T., Mosser, B., Hekker, S., et al. 2010, A&A, 522, A1
Karoff, C., Campante, T. L., Ballot, J., et al. 2013, ApJ, 767, 34
Kiefer, R. 2018, PhD Thesis, Kiepenheuer Institute for Solar Physics, Freiburg
Kurtz, D. W., Saio, H., Takata, M., et al. 2014, MNRAS, 444, 102
Lamb, H. 1932, Hydrodynamics (New York: Dover)
Lenz, P., & Breger, M. 2014, Period04: Statistical analysis of large Astronomical time Series, Astrophysics Source Code Library, ascl:1407.009
Mathur, S., Garcia, R. A., Regulo, C., et al. 2010, arXiv:1003.4749
Mazumdar, A. 2005, A&A, 441, 1079
Metcalfe, T. S., Creevey, O. L., Doğan, G., et al. 2014, ApJS, 214, 27
Monteiro, M. J. P. F. G. 2008, ApS&S, 316, 121
Mosser, B., & Appourchaux, T. 2009, A&A, 508, 877
Moya, A., & Garrido, R. 2008, ApS&S, 316, 129
Nielsen, M. B., Davies, G. R., Chaplin, W. J., et al. 2023, A&A, 676, A117
Nielsen, M. B., Ball, W. H., Standing, M. R., et al. 2020, A&A, 641, A25
Paxton, B., Smolec, R., Schwab, J., et al. 2019, ApJS, 243, 10
Provost, J. 2008, ApS&S, 316, 135
Ren, Y., & Jiang, B.-W. 2020, ApJ, 898, 24
Ren, Y., Zhang, Z., Jiang, B., Soszyński, I., & Jayasinghe, T. 2024, IAU Symp. 376: At the crossroads of astrophysics and cosmology: Period-luminosity relations in the 2020s ed. P. de Grijs, A. Whitelock, & M. Catelan (Cambridge: Cambridge Univ. Press) 195
Roxburgh, I. W. 2008, ApS&S, 316, 141
Roxburgh, I. W., & Vorontsov, S. V. 2006, MNRAS, 369, 1491
Scuflaire, R., Montalbán, J., Théado, S., et al. 2008, ApS&S, 316, 149
Smolec, R., & Moskalik, P. 2008, AcA, 58, 193
Sperl, M. 1998, CoAst, 111, 1
Sreenivas, K. R., Bedding, T. R., Li, Y., et al. 2024, MNRAS, 530, 3477
Stello, D., Chaplin, W. J., Basu, S., Elsworth, Y., & Bedding, T. R. 2009, MNRAS, 400, L80
Stello, D., Zinn, J., Elsworth, Y., et al. 2017, ApJ, 835, 83
Suran, M. D. 2008, ApS&S, 316, 163
Townsend, R. H. D., & Teitler, S. A. 2013, MNRAS, 435, 3406
Ulrich, R. K. 1986, ApJL, 306, L37
Viani, L., Basu, S., Corsaro, E., Ball, W. H., & Chaplin, W. J. 2019, ApJ, 879, 33
Zinn, J. C., Stello, D., Huber, D., & Sharma, S. 2019, ApJ, 884, 107

AAS | IOP Astronomy

Asteroseismology for the Nonspecialist

Derek L Buzasi

Chapter 7

Uncertainties and Limitations

Measured frequencies, amplitudes, and phases are always subject to uncertainty, and that uncertainty of course maps itself onto the physical properties of interest to us, such as mass, radius, and age, as well as internal structural information we derive from grid-based and other modeling of the individual frequencies themselves. However, there are other contributors to the final uncertainties, including

- Different methods for calculating global seismic parameters: As discussed in Section 6.2, methods for calculating both $\Delta\nu$ and $\nu_{\max}$ (as well as other values such as $\delta\nu$ and granulation parameters) vary, which leads to variation in the derived values of these parameters.
- Uncertainty in reference values: When we make use of the scaling relations, the scaling that takes place is relative to *solar* values, so uncertainty in those maps directly to physical uncertainties. While we know the solar mass, radius, and effective temperature to great precision, the same is not true for solar values of $\nu_{\max}$ and the large separation $\Delta\nu$.
- Uncertainty in the scaling relation: The exponents in the scaling relations given in the previous chapter derive from physical arguments. However, in practice the relations, while perhaps surprisingly good across a wide range of stellar physical parameters, are not perfect, and this is another source of uncertainty.
- Choice of stellar model: Table 1.1 lists a range of stellar interiors and evolution codes, each of which involves modestly different choices for input physics as well as computational algorithms. Choosing a different stellar model will invariably produce different oscillation frequencies; while these will differ only modestly (we hope!), they are nonetheless a source of uncertainty.
- Choice of model inputs: Input composition, choice of model integration grid and timestep, and numerous other choices also impact the output frequencies and can be considered to contribute to uncertainty inasmuch as a different researcher, even using the same code, is likely to arrive at slightly different values.

doi:10.1088/2514-3433/ae03a0ch7

We will start by looking again at the scaling relations, then move on to measurement uncertainties for parameters measured from the power spectrum, and end with a brief discussion of the impacts of choices of models and inputs.

Two important facts to always keep in mind about any real time series are that it contains noise, and that it is discretely sampled for a finite period of time. Furthermore, the noise is frequently not white, but generally increases toward lower frequencies, and the time series itself at best contains gaps, and at worst is fully nonuniformly sampled, particularly in the case of ground-based data.

7.1 Global Asteroseismic Parameters: Scaling Relations

Let's consider the frequency of peak power $\nu_{\rm max}$ and the mean large separation $\Delta\nu$ as global seismic parameters. As we have seen, they have the advantage of being measurable even in a fairly noisy power spectrum, and lead to the scaling relations introduced in Section 7.1:

$$
\begin{aligned}
\frac{R}{R_\odot} &= \left(\frac{\Delta\nu_{\rm max}}{\nu_{\rm max,\odot}}\right)\left(\frac{\Delta\nu}{\Delta\nu_\odot}\right)^{-2}\left(\frac{T_{\rm e}}{T_{\rm e,\odot}}\right)^{0.5} \\
\frac{M}{M_\odot} &= \left(\frac{\Delta\nu_{\rm max}}{\nu_{\rm max,\odot}}\right)^{3}\left(\frac{\Delta\nu}{\Delta\nu_\odot}\right)^{-4}\left(\frac{T_{\rm e}}{T_{\rm e,\odot}}\right)^{1.5} \\
\frac{\rho}{\rho_\odot} &= \left(\frac{\Delta\nu}{\Delta\nu_\odot}\right)^{2}\left(\frac{T_{\rm e}}{T_{\rm e,\odot}}\right)
\end{aligned}
\tag{7.1}
$$

As usual, the values subscripted with $\odot$ correspond to the solar values, so that these global seismic parameters can be immediately related to *physical* parameters of interest.

Despite the seductive simplicity of using relationships 7.1, there are proper concerns. To start with, as we've seen, measuring the values of $\nu_{\rm max}$ and $\Delta\nu$ is not always a straightforward exercise. The oscillation peaks lie on top of a noise floor, which has a more complex structure than simple white noise. Noise due to spacecraft pointing jitter and other instrumental sources generally manifests as red, so the impact of noise depends (potentially, a lot) on if the p-mode peak lies at low or high frequencies.

While power due to granulation and supergranulation is generally characterized by a Harvey function (Harvey 1985), and can be fit by three or more Harvey terms in addition to a white noise floor (with the number of terms potentially depending on data quality and time series length as well as the characteristics of the star itself), a more honest characterization includes an additional term:

$$
P(\nu) = \eta^2(\nu)\sum_{i=1}^{N}\frac{\xi_i\sigma_i^2\tau_i}{1 + (2\pi\nu\tau_i)^{a_i}}
\tag{7.2}
$$

Here $\eta(\nu) \leqslant 1$ is the apodization of signal amplitude as a function of frequency due to the finite sampling interval (Chaplin et al. 2011), while as before τ_i, σ_i, and a_i, respectively, give the characteristic timescale, σ, and decay slope of each background component. Overlying this multicomponent background lies the p-mode bump, most frequently fit with a Gaussian envelope of the form

$$G(\nu) = P_g \exp\left[-\frac{(\nu - \nu_{\max})^2}{2\sigma_g^2}\right] \tag{7.3}$$

Here P_g is the maximum p-mode power, $\nu_{\max}$ is the frequency of peak power, and σ_g characterizes the width of the envelope.

The details of the fitting procedure vary significantly from user to user, however. To make convergence easier, the power spectrum is generally smoothed, and different authors use different smoothing types (e.g., boxcar, Gaussian, etc.) and criteria (e.g., boxcar or Gaussian width); even if no smoothing is done, not all authors use the same frequency grid in their calculated power spectra. The number of Harvey components N chosen varies from 1 to 3, as does the range of allowed values for the decay parameters a_i. Finally, though many authors fit a Gaussian to the p-mode peak, others simply take the peak of the (usually heavily) smoothed power spectrum, or the first moment of the area under the bump. In addition, while for cool main sequence stars the locations in frequency space of the p-mode bump and the characteristic frequencies of granulation are well-separated and thus relatively straightforward to fit without crosstalk, this becomes less true near and on the giant branch, so the shape of the p-mode bump here can be much more sensitive to the specific granulation model chosen.

The second global component is the mean large separation $\Delta\nu$. The most straightforward way to measure this is to peak-bag the power spectrum and determine the mean separation directly from the measured frequency peaks. However, this is computationally demanding when large numbers of stars are involved, and may not even be possible in low signal-to-noise instances. Alternative approaches fall into two broad categories, based either on the power spectrum and so carried out in the frequency domain, or on the time series itself, and thus in the time domain. Generally, in the first case, one searches for periodicities in the power spectrum, using either an autocorrelation function or by taking the power spectrum of the power spectrum. Alternatively, other methods perform an autocorrelation in the time domain. In general, the signal strength from $\Delta\nu$ is low enough that performing an autocorrelation or $PS \otimes PS$ over the entire power spectrum would result in extremely low signal-to-noise for the detection peak, so the range of frequencies used is limited to a region around $\nu_{\max}$; how that region is defined varies from application to application, as does the frequency resolution of the power spectrum itself.

In addition to the uncertainties arising from the various algorithmic choices described above, additional uncertainties are created from the use of the scaling relations themselves. First, in addition to $\Delta\nu$ and $\nu_{\max}$, derived physical quantities such as mass M and radius R also depend on T_e, which in turn is typically derived from either spectroscopy or photometry, and tied to a bolometric luminosity standards which in turn derive from parallax and interferometric measurements (see Chapter 1). Each of these steps contributes uncertainty to the effective temperature and in turn to the mass, radius, and density estimates. Second, the scaling relations make use of the values of the ν_{max} and $\Delta\nu$ relative to their solar values, but these are not known exactly, and different authors adopt different values for them (Hekker 2020). While the range of variation in $\Delta\nu_{\odot}$ is relatively small, ranging from 134.88 μHz to 135.5 μHz, the range in ν_{max} is considerably larger (3050 to 3166 μHz; see Table 7.1). The corresponding relative uncertainties (based on the standard deviation of the distribution of values) are 0.1% and 1.8%, which would translate to uncertainties in radius and mass of 2% and 5.8%, assuming everything else is known perfectly (Figure 7.1).

Finally, though the scaling relations appear to be model-independent, in fact this is really true only to first order. Comparisons to results from models based on complete frequency lists rather than to global parameters, and to results derived from other techniques such as classical eclipsing binary analysis, show that a more accurate version of the scaling relations shown in Equations (7.1) contain additional multiplicative terms $f_{\nu_{max}}$ and $f_{\Delta\nu}$:

$$\frac{R}{R_{\odot}} = \left(\frac{\Delta\nu_{max}}{f_{\nu_{max}}\,\nu_{max,\odot}}\right)\left(\frac{\Delta\nu}{f_{\Delta\nu}\,\Delta\nu_{\odot}}\right)^{-2}\left(\frac{T_e}{T_{e,\odot}}\right)^{0.5} \tag{7.4}$$

$$\frac{M}{M_{\odot}} = \left(\frac{\Delta\nu_{max}}{f_{\nu_{max}}\,\nu_{max,\odot}}\right)^{3}\left(\frac{\Delta\nu}{f_{\Delta\nu}\,\Delta\nu_{\odot}}\right)^{-4}\left(\frac{T_e}{T_{e,\odot}}\right)^{1.5} \tag{7.5}$$

$$\frac{\rho}{\rho_{\odot}} = \left(\frac{\Delta\nu}{f_{\Delta\nu}\,\Delta\nu_{\odot}}\right)^{2}\left(\frac{T_e}{T_{e,\odot}}\right) \tag{7.6}$$

Table 7.1. The Range of Solar Reference Values Used in the Literature for the Global Asteroseismic Parameters ν_{max} and $\Delta\nu$.

$\Delta\nu$ (μHz)	ν_{max} (μHz)	Source
134.9	3050	Kjeldsen & Bedding (1995)
134.88	3120 ±5	CAN; Kallinger et al. (2010)
134.9	3150	Chaplin et al. (2011)
135.1 ±0.1	3090 ±30	Huber et al. (2011)
135.5	3050	Mosser et al. (2013)
134.9	3060 ±10	COR; Hekker et al. (2011)
135.03 ±0.07	3140 ±13	OCT; Hekker et al. (2011)
134.88 ±0.03	3140 ±5	Kallinger et al. (2014)
135.4 ±0.3	3166 ±6	Themeßl et al. (2018)
135	3050	A2Z; Mathur et al. (2010)

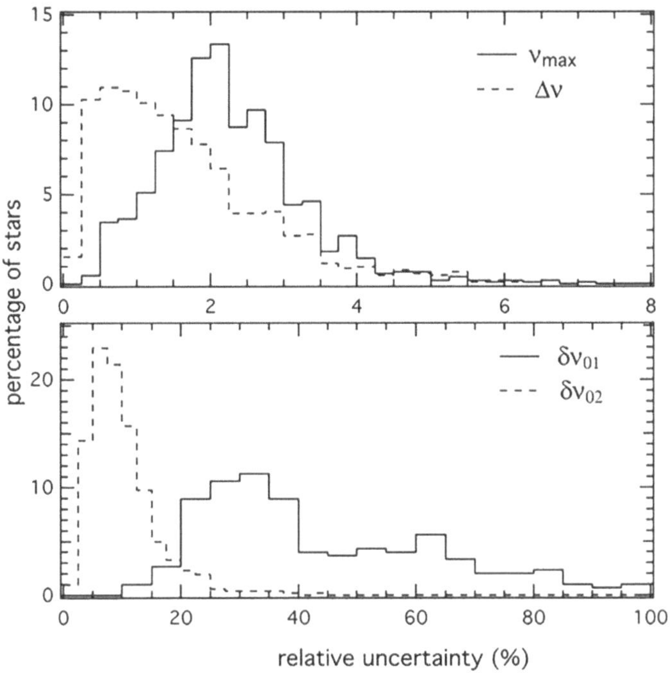

Figure 7.1. Histograms of the relative uncertainties for ν_{max}, the large frequency separation $\Delta\nu$, and the small separations $\delta\nu_{01}$ and $\delta\nu_{02}$. Reprinted with permission from Kallinger et al. (2010)

Sharma et al. (2016) and Sharma et al. (2017) derived these functions using the stellar evolution code MESA combined with the oscillation code GYRE (see also Rodrigues et al. 2017 and Tayar 2024) for a variety of points along multiple evolutionary tracks, and Figure 7.2 illustrates the results. For even modestly evolved stars, the correction factors are typically 2% and can be as large as 4%.

Some efforts have been made to improve the scaling relations. Bellinger (2019) noted that published scaling relations could be generalized to a form

$$\frac{Y}{Y_\odot} = \left(\frac{\nu_{max}}{\nu_{max,\odot}}\right)^\alpha \left(\frac{\Delta\nu}{\Delta\nu_\odot}\right)^\beta \left(\frac{\delta\nu}{\delta\nu_\odot}\right)^\gamma \left(\frac{T_e}{T_{e,\odot}}\right)^\delta \exp([\mathrm{Fe/H}])^\varepsilon \tag{7.7}$$

where Y represents some physical parameter of interest such as mass or radius, and $\delta\nu$ the mean small frequency separation. Bellinger (2019) then empirically calibrated the relationship using asteroseismic data for 80 stars with the best model fits available. The results are shown in Table 7.2, with variables labeled as appropriate for Equation (7.7). A few points of note:

- The effect of the small separation $\delta\nu$ on the results is minimal, which is welcome because that parameter is frequently unknown for stars without full frequency solutions, and can be noisy even when it is known. Accordingly, Bellinger (2019) also derived scaling relations for mass M and radius R without the small frequency separation, and argues that even though

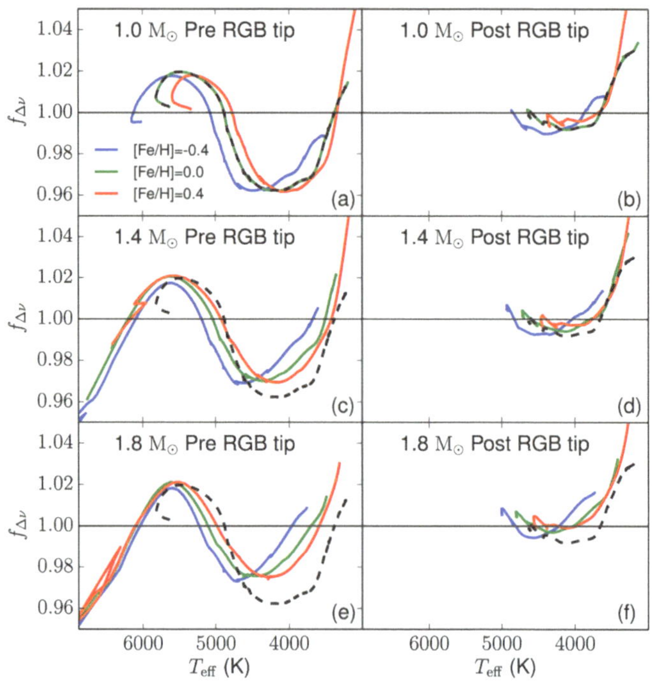

Figure 7.2. Correction factor $f_{\Delta\nu}$ predicted by stellar models as a function of T_e for a range of metallicities, masses, and evolutionary states. The left panels describe evolution from the main sequence to the tip of the RGB, while the right panels apply to evolution from the onset to the end of He core-burning. The dashed reference line is for the case with [Fe/H] = 0 (for $Z_\odot = 0.019$) and $M = 1M_\odot$. Reprinted with permission from Sharma et al. (2016).

Table 7.2. Classical and New MCMC-Fitted Exponents for Scaling Relations: From Bellinger (2019).

	Y	$\nu_{\rm max}$ α	$\Delta\nu$ β	$\delta\nu$ γ	T_e δ	[Fe/H] ε
Classic	M	3	−4	–	1.5	–
New	M	0.975	−1.435	–	1.216	0.270
Classic	R	1	−2	–	0.5	–
New	R	0.305	−1.129	–	0.312	0.100
Seismic	R	0.883	−1.859	–	–	–
New	τ	−6.556	9.059	−1.292	−4.245	−0.426

neglecting this dependence decreases the fit quality, the resulting scaling relations are still superior to the older ones based on Kjeldsen & Bedding (1995). These updated scaling relations do rely on metallicity measurements [Fe/H], which may not be available, or which may not be of high quality, for all stars.

- Bellinger (2019) estimated the uncertainties resulting from application of these relations using a jackknife procedure. The resulting scaling relations produce uncertainties on $\delta M/M$ and $\delta R/R$, which are of order 1% in mass and 0.3% in radius.
- The procedure also leads to a scaling relation for stellar *age* for the first time. In this case, the small frequency separation cannot be neglected due to its known age sensitivity. Typical relative uncertainties are ~10%, but tend to decrease with increasing age. While the mass and radius estimates from these revised scaling relations appear unbiased, there are some biases in the age relation, so it should be used with more care.
- The exponents in the original scaling relations were based on physical arguments, while those in Bellinger (2019) are derived based on purely empirical considerations. This may be part of the reason that they have not been universally adopted to date. Furthermore, the 80 stars used to derive the relationship span roughly $M = [0.8, 1.5]\,M_\odot$, $R = [0.8, 2.5]\,R_\odot$, and $\tau = [1, 11]$ Gyr, so the fits may be unreliable outside of these ranges. However, note that Bellinger (2020) extended the new scaling relations to giant stars.

7.2 Choosing Significance Criteria

Our discussion of periodogram uncertainties is necessarily somewhat limited. For further information and deeper insights, sources abound, starting with Lomb (1976); Scargle (1982); Horne & Baliunas (1986); Schwarzenberg-Czerny (1998a, 2003); Frescura et al. (2008), and Baluev (2008).

Figure 7.3 shows a sample time series and its resulting power spectrum. The light curve is 12.7 days long (half a TESS sector), sampled at the TESS short-cadence 20 s intervals, with white noise added at a level of $\sigma = 10^{-3}$ of the mean flux level; only a short segment of the full time series is shown. Two artificial oscillations have also been added to the time series. A piece of the power spectrum is shown; note that in practice different implementations make use of different scalings for the power spectrum. The default here is the *power spectral density* (V^2 Hz^{-1} or ppm^2 μHz^{-1} or the like), but simple squared magnitude, or V^2 is also an option; different groups, studying different types of stars, have different standards. The peak in the periodogram certainly stands out. But there are other peaks that stand out as well. How do we know which one is (or are!) significant, statistically speaking? What criteria can we apply to determine if the peak is significant? There are a number of options.

First, we can take advantage of a fundamental statistical property of the periodogram, which is that the values $P(f)$ are exponentially distributed, **if** the time series is evenly spaced and the underlying noise is Gaussian. This helps us in theory because it allows us to estimate the probability that a peak of a certain size in the periodogram will occur by chance.

The basis for these tests is the single trial cumulative distribution function (CDF),

$$F_Z(z) = [Z \leqslant z] \tag{7.8}$$

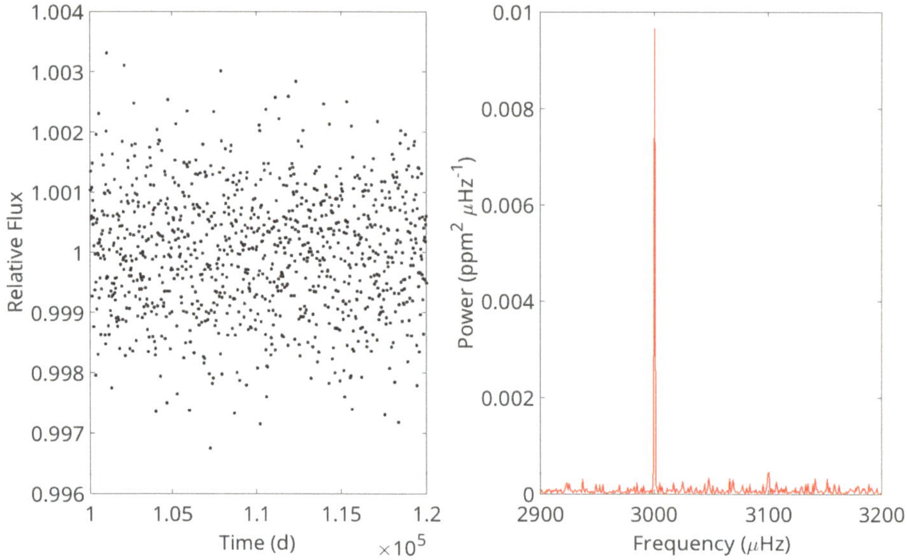

Figure 7.3. On the left, a time series containing two oscillation frequencies, generated at the TESS short-cadence 20 s rate with white noise added at a level of $\sigma = 10^{-3}$ of the mean flux level; only a short segment of the full time series is shown. The right-hand panel shows a portion of the power spectrum for the time series.

Here the random variable $Z = P_X(\omega)$ is the periodogram power at frequency ω for the time series X, and z is some specified power threshold. The function $P_Z(z)$ gives the probability that, when the data X are pure white noise, the periodogram power at the given frequency ω is below z. We can use this CDF to test for significance as follows. Suppose a model predicts an oscillation at frequency ω, so we expect $P_X(\omega)$ to be large at that frequency. However, pure noise could randomly produce a peak that large at that frequency.

The probability that $P_X(\omega)$ for some given threshold z_0 is $p_0 = F_Z(z_0)$. Inverting this function, we obtain the threshold power level z_0 for which a power value $Z \leqslant z_0$ has a probability p_0 of being due to pure noise. It is given by $z_0 = Fz^{-1}(p_0)$. Equivalently, a power value $P_X(\omega) > z_0$ has a probability $1 - p_0$ of being due to pure noise.

Things are more complex though, because we don't actually evaluate the periodogram at a single frequency, but at some selected set ω_i. We then plot the result and search for peaks. Suppose the data are pure noise. To calculate the probability that **all** of the periodogram powers at **all** of the sampled frequencies fall below some threshold z, we define a new random variable

$$Z_{max} = \max[P_X(\omega_i)] \tag{7.9}$$

Z_{max} is the maximum periodogram power among all N sampled powers. Now, the power at each of the sample values will fall below z if $Z_{\max} \leqslant z$, so we need to calculate the CDF

$$F_{Z_{\max}}(z) = Pr[Z_{\max} \leqslant z] \tag{7.10}$$

This function gives the probability that the periodogram power doesn't rise above the threshold z at ANY of the sampled frequency. We construct a new significance test, with z_0 as the specified power threshold. Now the probability that pure noise alone will produce a set of $P_X(\omega_i)$ that never exceed z_0 is given by

$$p_0 - F_{Z_{\max}}(z_0) \tag{7.11}$$

which inverts to give

$$z_0 = F^{-1}_{Z_{\max}}(p_0) \tag{7.12}$$

If the periodogram values under Gaussian noise are exponentially distributed, then

$$p_Z(z)dz = Pr[z < Z < z + dz] = \frac{1}{\sigma_X^2} e^{-z/\sigma_X^2} dz \tag{7.13}$$

The CDF then becomes

$$P_Z(z) = Pr[Z < z] = \int_{\psi_0}^{z} p_Z(\psi)d\psi = 1 - e^{-z}/\sigma_X^2 \tag{7.14}$$

We are interested in the probability that the periodogram power at our frequency of interest is greater than our specified threshold z, which then is

$$Pr[Z > z] = 1 - P_Z(z) = e^{-z/\sigma_X^2} \tag{7.15}$$

Suppose now that we evaluate the periodogram at frequencies ω_i. Denote the periodogram powers at these frequencies by $Z_i = P_X(\omega_i)$. If the data X are pure noise, then the Z_i are random variables. We are interested in determining the probability that this entire set of observed periodogram powers could have been produced by pure noise alone. To make the problem tractable, we assume that the Z_i are independent random variables (Scargle 1982). Then, to calculate the probability that all the sampled periodogram powers are less than some specified threshold power z, define a new random variable

$$Z_{\max} = \max(Z_i) \tag{7.16}$$

and the probability that any *particular* power Z_A in this set falls below the threshold is

$$Pr[Z_A < z] = 1 - e^{-z/\sigma_X^2} \tag{7.17}$$

and the probability that *all* of the Z_i do is

$$Pr[Z_i < z] = \left[1 - e^{-z/\sigma_X^2}\right]^N \tag{7.18}$$

where we can make the last step by assuming that the Z_i are all independent. This means that the probability that *at least one* of the peaks is above the threshold z is

$$Pr[Z_i > z] = 1 - \left[1 - e^{-z/\sigma_X^2}\right]^N \tag{7.19}$$

We can adopt this as our false alarm probability (FAP).

What does it mean to have a set of independent frequencies? If the time series is evenly spaced, that it can be shown (Scargle 1982) that there is a set of frequencies at which the power can be treated as independent variables. These are the "natural frequencies," well-named because they are constructed from the parameters of the time series itself, so for a time series with N evenly spaced points, there are $N/2$ members of the corresponding set of natural frequencies, evenly spaced by

$$\Delta\omega = \frac{2\pi}{T} \tag{7.20}$$

or

$$\Delta\nu = \frac{1}{T} \tag{7.21}$$

where T is the total length (in time units) of the time series and k is the number of points. The lowest frequency, $\omega = \nu = 0$, is just the DC term, and generally the mean value is subtracted from the time series before processing.[1]

If data are *unevenly* sampled (a case which of course also applies to data with gaps) life is more complicated. In that case, there is no Nyquist frequency and the periodogram of noise is no longer guaranteed to be exponentially distributed. In this case, Scargle (1982) suggests continuing to use the natural frequencies that would be appropriate for an evenly spaced time series with the same number of points N and total length T.

Koen (1990) and Schwarzenberg-Czerny (1998a, 1998b) have pointed out that the assumption that we *know* the variance σ_X^2 of the time series is incorrect, and that this means that we only estimate it, which leads to a more complex probability estimate for $Z_{\max} > z$:

$$Pr[Z_{\max} > z] = 1 - \left[1 - \left(1 - \frac{z/\sigma_X^2}{[N/2]}\right)^{N_i/2}\right]^N \tag{7.22}$$

Here N is the number of points in the periodogram and N_i is the number of mutually independent frequencies. In the limit where $N \to \infty$, this estimate reduces to the simpler one in Equation (7.19), and even in more modest (realistic) time series, the difference is generally not significant enough to be meaningful.

However, this approach does not avoid the fundamental issue, which is that when the data are unevenly spaced a set of natural frequencies simply no longer exists! Formally, this leads to the conclusion that in these circumstances we should perform Monte Carlo simulations in all cases in order to estimate peak significance, which is potentially time-consuming and may incur additional problems. In reality, many practitioners simply adopt the Scargle (1982) Equation (7.19), with the number N_i of

[1] Practically speaking, doing this also avoids the potentially huge DC peak (at zero frequency), which can visually swamp other structure in the periodogram.

independent frequencies given by Equation (7.21). Numerical simulations using Monte Carlo techniques do appear to indicate that, for large time series, this approach at worst underestimates the significance of individual peaks, so it has the virtue of being conservative. A further difficulty is that the underlying noise profile is not generally white, whether that is due to contributions from stellar granulation and super-granulation or from instrumental effects such as pointing jitter or thermal drifts. If the frequencies of interest are high enough that the *local* noise is essentially white, this may not matter, but (as usual) there is no substitute for simply making a log-log plot of the periodogram and visually inspecting your data to make sure this is the case!

Another common approach involves simply calculating the signal-to-noise ratio (SNR) of the individual peaks, typically relative to the immediately local region of the power spectrum, and taking peaks as significant that exceed some particular SNR, usually $SNR > 4$ (roughly). In the presence of white noise, adopting SNR = 4 is roughly equivalent to a false alarm probability of 2% at any given frequency (assuming the usual TESS single-sector data sets), so this is not on its face an unreasonable approach. It also has the advantage of being easily adapted to the presence of non-white noise, by the simple expedient of using the local noise level in the power spectrum to calculate the peak SNR. However, it has little or no theoretical support, so should in most cases be backed by an alternate approach.

One approach to constructing Monte Carlo simulations to test significance is to generate time series using the times t_i from the original time series and the measured peak(s) amplitude, frequency, and phase. To these time series we then add realizations of noise based either on the Gaussian noise estimated from the original data set, or the non-white noise based on a fit to the underlying noise visible in the power spectrum. The resulting parameters are then measured for each realization and uncertainties inferred from their distribution.

Finally, another approach to estimating significance is to simply fit a sinusoid in a narrow window of both amplitude and frequency, derived from the periodogram information, obtain uncertainty estimates from the fit, and use those estimates to infer significance. This has the advantage of potentially obtaining a more accurate frequency fit than is possible from the frequency grid used to calculate the periodogram, making use of the existing noise characteristics in the time series, and also simultaneously supplying uncertainties in the other fit parameters (frequency, amplitude, phase). However, particularly in cases where numerous oscillation frequencies are present in close proximity, this does not address numerical interactions between the different frequencies. It is possible to add more sinusoids to the fit function to address this issue, but this can increase complexity rapidly and to the point where simply adopting the Monte Carlo approach is preferable.

As a reminder, the user is free to choose the frequencies at which the periodogram is evaluated, but there are important considerations to keep in mind. Generally, evaluation is done on a regularly-spaced frequency grid, which can then be characterized by three parameters

- Minimum frequency $f_{\rm min}$: This typically is set to either 0 or the inverse of the length of the time series T, which represents the longest sinusoid for which a complete period can fit within the time series.

- Maximum frequency f_{max}: This is the shortest period which can be fully sampled by the time series. When the time series is evenly sampled by some Δt, the maximum frequency is the Nyquist frequency, $2/\Delta t$.
- Frequency spacing Δf: For an evenly sampled time series, the minimum typical $\Delta f = 1/T$. However, this risks missing narrow peaks in the periodogram which might fall between the frequency grid points, underestimating their amplitude. Accordingly, the power spectrum is usually *oversampled*, typically by up to a factor of ~ 8.

Note that, in cases where the frequency or frequencies of interest are known *a priori*, computation time can be saved by adjusting f_{min} and f_{max} to include only that area of interest. However, this obviously runs the risk of missing any previously unknown oscillation frequencies, and may also impact the ability to properly and fully characterize the underlying noise in the time series. Note also that oversampling the periodogram does NOT increase the number of independent frequencies!

7.3 Estimating Uncertainties in Measured Parameters

We generally measure specific peaks in the power spectrum or periodogram, which results in values for frequency, amplitude, and phase. If that measurement process is considered as the fitting of a sinusoid to the data (or, equivalently, performing a Lomb–Scargle periodogram), and the noise is white, then there is a formal solution for the resulting measurement uncertainties.

Take a typical asteroseismic time series, which consists of N measurements of the magnitudes or fluxes (or, indeed, velocities), which we label as $m_i \pm \Delta m_i$, where the measurement uncertainties are random and normally distributed. These measurements, whatever they are, are made at discrete times t_i, and we assume that the uncertainties in the time measurements are negligible, generally a good assumption. Let's suppose that our data set contains only one oscillation frequency, so, when we search for frequencies in this time series, we are effectively fitting a sinusoid to it, so

$$f_i(t_i) = A_0 + A\sin(\omega t_i + \phi) \tag{7.23}$$

Here A_0 is the mean value of the time series (the DC component), and we take ω as known (from, for example, prior observations),[2] so we are really searching for the amplitude A and phase factor ϕ of the sinusoid.

To fit the sinusoid, we perform χ^2 minimization, where χ^2 is defined in the usual way as

$$\chi^2 = \sum_{i=1}^{N}[m_i - f(t_i)]^2 = \sum_{i=1}^{N}[m_i - A_0 + A\sin(\omega t_i + \phi)]^2 \tag{7.24}$$

[2] An example of this might be when one has a long ground-based observation history giving an extremely precise frequency, and is using that frequency to analyze a single TESS sector of observations, which necessarily has poor frequency resolution due to its short length.

Minimizing with respect to the three variables of interest, A_0, A, and ϕ produces

$$\begin{aligned}\frac{\partial\chi^2}{\partial A_0} &= 0\\ \frac{\partial\chi^2}{\partial A} &= 0\\ \frac{\partial\chi^2}{\partial\phi} &= 0\end{aligned} \tag{7.25}$$

Solving gives

$$\begin{aligned}\frac{1}{N}\sum_{i=1}^{N} m_i &= A_0\\ \frac{2}{N}\sum_{i=1}^{N} m_i \sin(\omega t_i + \phi) &= A\\ \sum_{i=1}^{N} m_i \cos(\omega t_i + \phi) &= 0\end{aligned} \tag{7.26}$$

Here we've assumed that the time series are sufficiently long that the continuous identities $\overline{\cos^2 x} = 1/2$ and $\overline{\sin x \cos x} = 0$ are satisfied in the discrete case as well. Next we take a total differential of Equation (7.26)b to get

$$\delta A = \frac{2}{N}\sum_{i=1}^{N}[\delta m_i \sin(\omega t_i + \phi) + m_i \cos(\omega t_i + \phi)\delta\phi] = \frac{2}{N}\sum_{i=1}^{N}\delta m_i(\sin\omega t_i + \phi) \tag{7.27}$$

where we've made use of Equation (7.26)c to remove the second term. Now we take the root-mean-square of this quantity, so that

$$\begin{aligned}\overline{(\delta a)^2} &= \frac{4}{N^2}\sum_{i=1}^{N}\sum_{j=1}^{N}\delta m_i \delta m_j \sin(\omega t_i + \phi)\sin(\omega t_j + \phi)\\ &= \frac{4}{N^2}\sum_{i=1}^{N}(\delta m_i)^2 \sin^2(\omega t_i + \phi)\\ &= \frac{4\sigma^2}{N^2}\sum_{i=1}^{N}\sin^2(\omega t_i + \phi)\\ &= \frac{2\sigma^2}{N}\end{aligned} \tag{7.28}$$

Since $\overline{(\delta a)^2} = \sigma_A^2$, we have

$$\sigma_A = \sigma\sqrt{\frac{2}{N}} \tag{7.29}$$

Unsurprisingly, the amplitude estimate improves with more points in the time series. We can do similar derivations for the other parameters, and find

$$\sigma_f = \sqrt{\frac{6}{N}}\frac{1}{\pi T}\frac{\sigma}{A} \tag{7.30}$$

$$\sigma_\phi = \frac{1}{2\pi}\sqrt{\frac{2}{N}}\frac{\sigma}{A} \tag{7.31}$$

These formulae are commonly used to calculate uncertainties in the published literature, but we should be careful! Their derivation assumed white noise, and applied to the measurement of a single frequency and not a prewhitened time series. As such, these values represent the very best we can do, and almost always will overstate the precision possible. In the end, Monte Carlo simulations are the gold standard.

7.4 Surface Effect

Many approximations in both stellar interiors modeling, such as the adiabatic approximation, temperature gradients and mixing-length models for convection, fail as we approach the stellar surface (or more accurately the $\tau = 1$ depth). Simultaneously, frequently-neglected physical processes such as magnetic fields and dynamic pressure become relatively more important in that region. These factors manifest themselves in systematic frequency shifts, which can be quite large (see Figure 6.8), particularly at higher frequencies where the near-surface regions contribute more substantially. Accordingly, a correction for the surface effect must be made, especially if the higher-frequency modes are to be useful. The effect varies as a function of mode inertia I_{nl}, in the sense that modes with greater inertia are affected less by the surface effect, but after correcting for this effect by scaling each mode frequency by the corresponding I_{nl} an overall pattern remains which must be accounted for (Christensen-Dalsgaard & Berthomieu 1991).

There are several approaches to performing a surface correction. The most straightforward is to simply take the well-characterized solar correction and scale it to the star in question. However, it's not clear how the correction depends on mass, evolution state, and metallicity. The scaled-solar approach can be modestly improved by adding a linear offset to the fit,

$$\nu_{\rm corr} = \nu_{\rm mod} + a\delta\nu_{\odot,\rm SC} + b \tag{7.32}$$

where the b term is an attempt to mimic the effects of stellar evolution (White et al. 2012). Kjeldsen et al. (2008) have provided the most commonly used approach, which is to fit the correction with a power law of the form

$$\nu_{\rm corr} = \nu_{\rm model} + a\left(\frac{\nu_{\rm mod}}{\nu_{\rm ref}}\right)^b \tag{7.33}$$

Here $\nu_{\rm ref}$ is generally taken to be $\nu_{\rm max}$ and the exponents a and b are determined by fitting to the solar frequency differences. This approach tends to overcorrect at low frequencies, however, with potentially significant ramifications for models of the deeper interior. A number of other approaches of varying complexity are also extant in the literature (Roxburgh & Vorontsov 2003; Gruberbauer et al. 2013).

Most recently, Ball & Gizon (2014) have proposed an approach that does not depend on scaling from the solar surface correction. Their correction takes the form

$$\delta\nu = \frac{a_3}{I_{nl}}\left(\frac{\nu}{\nu_{\rm ac}}\right)^3 + \frac{a_{-1}}{I_{nl}}\left(\frac{\nu}{\nu_{\rm ac}}\right)^{-1}, \tag{7.34}$$

where I_{nl} is the mode inertia, though as a simplified alternative they suggest that the second term could be dropped. Schmitt & Basu (2015) compared several forms of the surface correction as applied to models, and found that the Ball & Gizon (2014) approach was superior, even for red giants.

Fortunately, Basu & Kinnane (2018) showed that the global parameters of a star (specifically mass, radius, and age) obtained from asteroseismology are robust to choice of surface term correction, though the internal structure may not be. Compton et al. (2018) agreed with this conclusion, once mode inertia scaling is included, and found that inclusion of the Ball & Gizon second term substantially improved fits for more evolved stars. Ong et al. (2021) noted that, while other global parameters were robust, the initial helium abundance Y_0 was sensitive to choice of method; they also found that nonparametric algorithms returned a smaller spread to inferred masses for red giants than did Ball & Gizon (2014), and Ong (2024) strengthened the argument that existing parametric surface correction approaches are insufficient for studying red giant rotation.

7.5 Estimating Uncertainties in Physical Parameters

Once significant peaks are identified in the power spectrum and their frequencies and uncertainties are established and corrected for the surface effect, our focus changes to characterizing the uncertainties in the stellar physical parameters derived from the frequencies.

One approach is something of a hybrid between application of the scaling relations and modeling of individual frequencies, and involves using a stellar evolution code to perform grid modeling and extracting global asteroseismic parameters such as $\nu_{\rm max}$, $\Delta\nu$, and ΔP from those grids, and interpolating within the grids for comparison to measured stellar values. Rodrigues et al. (2017) found that this approach recovered masses to a precision of 5% and ages to within 19%, and including luminosity measurements in the fit improved those values to 3% and 10%, based on their ability to recover model input values. Figure 7.4 illustrates the procedure.

These results are likely to represent something of a best-case scenario, however. Tayar & Joyce (2025) compared seismic ages, derived based on global parameters, to isochrone ages for a range of globular clusters, and found that the typical scatter in the asteroseismic ages was at least 5 times larger than the isochrone age uncertainties. Why the difference? One reason, as noted in Section 1.8, is that the fundamental underpinning of our understanding of basic stellar parameters is anchored by measurements of stellar diameters (via interferometry) and bolometric fluxes, and

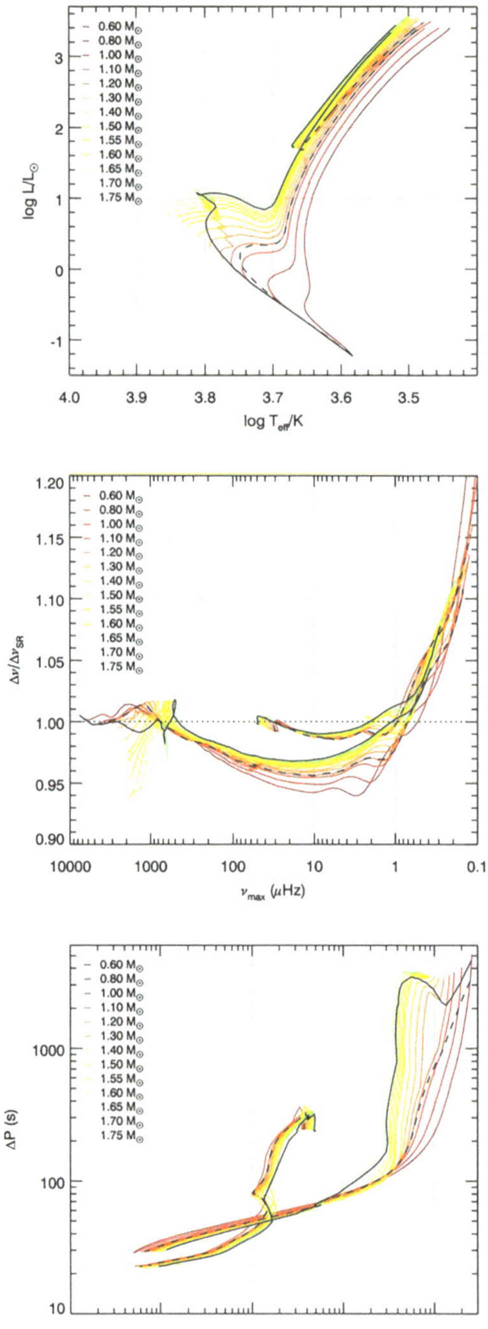

Figure 7.4. Top panel: Model evolutionary tracks shown on the HR diagram. Middle panel: The divergence of model large separation from that predicted by the scaling relation, $\Delta\nu/\Delta\nu_{SR}$, versus ν_{max}. Bottom: ΔP versus $\Delta\nu$. In each panel tracks are color-coded by mass, and the solid (dashed) line indicates an interpolated isochrone at an age of 2 (10) Gyr. Reprinted with permission from Rodrigues et al. (2017).

these imply **minimum** uncertainties of 2.4% on T_e, 2% on L, and 4.2% on radius (Tayar et al. 2022). In addition, differences in assumptions and input physics between different models lead to grid differences, which map to uncertainties in mass of order 5% and in age of closer to 20% (Silva Aguirre et al. 2020a, 2020b; Martins & Palacios 2013; Tayar et al. 2022).

Even within the space of a single modeling code, Li & Joyce (2025) found that simple changes in the adopted model resolution led to potentially large impacts on calculated values. The effect was higher for higher-mass and metallicity stars, and grew from under 1% for solar-like main sequence oscillators up to nearly 20% for AGB stars. Some of this uncertainty derives from our lack of knowledge of the surface correction term, so another approach involves defining parameters that are less sensitive to that correction. Roxburgh & Vorontsov (2003) found that ratios of the large and small frequency separation were nearly insensitive to the surface term

$$r_{l,l+2}(n) = \frac{\delta\nu_{l,l+2}(n)}{\Delta\nu_{1-l}(n+l)} \tag{7.35}$$

and

$$r_{l,1-l} = \frac{dd_{l,1-l}(n)}{\Delta\nu_{1-l}(n+l)} \tag{7.36}$$

where the various *dd* are the five-point averaged frequencies, so

$$dd_{0,1}(n) = \frac{1}{8}[\nu_0(n-1) - 4\nu_1(n-1) + 6\nu_0(n) - 4\nu_1(n) + \nu_0(n+1)] \tag{7.37}$$

and

$$dd_{1,0}(n) = \frac{1}{8}[\nu_1(n-1) - 4\nu_0(n) + 6\nu_1(n) - 4\nu_0(n+1) + \nu_1(n+1)] \tag{7.38}$$

Li & Joyce (2025) do in fact find that adopting the second-differences r_{02} and r_{13} defined by Roxburgh & Vorontsov (2003) is much more robust against changes in model resolution, perhaps a good reason to adopt these quantities.

Calculating full frequency fits can of course in principle do even better. Recent work tested the precision possible through the use of an eclipsing binary containing an oscillating red giant (Thomsen et al. 2025). In that case, the dynamically determined mass agreed with the pulsation mass to with 1.4%. This result is consistent with the goal of Baum et al. (2024) to use oscillating red giants in eclipsing binaries to obtain mass and radius to within 1% in order to better calibrate scaling relations. How stable these conclusions are across different modeling tools remains to be seen.

References

Ball, W. H., & Gizon, L. 2014, A&A, 568, A123
Baluev, R. V. 2008, MNRAS, 385, 1279
Basu, S., & Kinnane, A. 2018, ApJ, 869, 8

Baum, A., Pepper, J., & Hambleton, K. 2024, 8th TESS/15th Kepler Asteroseismic Science Consortium Workshop (Lisbon: Observatório Astronómico de Lisboa) 119
Bellinger, E. P. 2019, MNRAS, 486, 4612
Bellinger, E. P. 2020, MNRAS, 492, L50
Chaplin, W. J., Kjeldsen, H., Christensen-Dalsgaard, J., et al. 2011, Sci, 332, 213
Christensen-Dalsgaard, J., & Berthomieu, G. 1991, Solar Interior and Atmosphere, ed. A. N. Cox, W. C. Livingston, & M. S. Matthews (Tucson, AZ: Univ. Arizona Press) 401
Compton, D. L., Bedding, T. R., Ball, W. H., et al. 2018, MNRAS, 479, 4416
Frescura, F. A. M., Engelbrecht, C. A., & Frank, B. S. 2008, MNRAS, 388, 1693
Gruberbauer, M., Guenther, D. B., MacLeod, K., & Kallinger, T. 2013, MNRAS, 435, 242
Harvey, J. 1985, ESA Special Publication, Vol. 235, Future Missions in Solar, Heliospheric & Space Plasma Physics, ed. E. Rolfe, & B. Battrick (Noordwijk: ESA Scientific & Technical Publications) 199
Hekker, S. 2020, Frontiers in Astronomy and Space Sciences, 7, 3
Hekker, S., Elsworth, Y., De Ridder, J., et al. 2011, A&A, 525, A131
Horne, J. H., & Baliunas, S. L. 1986, ApJ, 302, 757
Huber, D., Bedding, T. R., Stello, D., et al. 2011, ApJ, 743, 143
Kallinger, T., Mosser, B., Hekker, S., et al. 2010, A&A, 522, A1
Kallinger, T., De Ridder, J., Hekker, S., et al. 2014, A&A, 570, A41
Kjeldsen, H., & Bedding, T. R. 1995, A&A, 293, 87
Kjeldsen, H., Bedding, T. R., & Christensen-Dalsgaard, J. 2008, ApJL, 683, L175
Koen, C. 1990, ApJ, 348, 700
Li, Y., & Joyce, M. 2025, arXiv:2501.13207
Lomb, N. R. 1976, ApS&S, 39, 447
Martins, F., & Palacios, A. 2013, A&A, 560, A16
Mathur, S., Garcia, R. A., Regulo, C., et al. 2010, arXiv:1003.4749
Mosser, B., Michel, E., Belkacem, K., et al. 2013, A&A, 550, A126
Ong, J. M. J. 2024, ApJ, 960, 2
Ong, J. M. J., Basu, S., & McKeever, J. M. 2021, ApJ, 906, 54
Rodrigues, T. S., Bossini, D., Miglio, A., et al. 2017, MNRAS, 467, 1433
Roxburgh, I. W., & Vorontsov, S. V. 2003, A&A, 411, 215
Scargle, J. D. 1982, ApJ, 263, 835
Schmitt, J. R., & Basu, S. 2015, ApJ, 808, 123
Schwarzenberg-Czerny, A. 1998a, BaltA, 7, 43
Schwarzenberg-Czerny, A. 1998b, MNRAS, 301, 831
Schwarzenberg-Czerny, A. 2003, ASP Conf. Ser. 292, Interplay of Periodic, Cyclic and Stochastic Variability in Selected Areas of the H-R Diagram ed. C. Sterken (San Francisco, CA: ASP) 383
Sharma, S., Stello, D., Bland-Hawthorn, J., Huber, D., & Bedding, T. R. 2016, ApJ, 822, 15
Sharma, S., Stello, D., Huber, D., Bland-Hawthorn, J., & Bedding, T. R. 2017, ApJ, 835, 163
Silva Aguirre, V., Christensen-Dalsgaard, J., Cassisi, S., et al. 2020a, A&A, 635, A164
Silva Aguirre, V., Christensen-Dalsgaard, J., Cassisi, S., et al. 2020b, A&A, 635, A164
Tayar, J. 2024, Cambridge Workshop on Cool Stars, Stellar Systems, and the Sun, Cambridge Workshop on Cool Stars, Stellar Systems, and the Sun (San Diego: San Diego State Univ.) 57
Tayar, J., Claytor, Z. R., Huber, D., & van Saders, J. 2022, ApJ, 927, 31

Tayar, J., & Joyce, M. 2025, ApJL, 984, L56
Themeßl, N., Hekker, S., Southworth, J., et al. 2018, MNRAS, 478, 4669
Thomsen, J. S., Miglio, A., Brogaard, K., et al. 2025, A&A, 699, A152
White, T. R., Bedding, T. R., Gruberbauer, M., et al. 2012, ApJL, 751, L36

Derek L Buzasi

Chapter 8

From Frequencies to Physics: An Introduction to Advanced Topics

Asteroseismology has found productive application to vast range of questions about stellar structure and evolution, and this chapter is not an attempt to do justice to all of those applications. Instead, I focus on a few areas that have significant community interest at present.[1]

8.1 Frequency Inversions and the Inverse Problem

8.1.1 An Example on A String

Let's go back to the vibrating string in Section 3.2, which is a useful paradigm for understanding how to use the oscillation frequencies to infer the internal structure of a star. We can rewrite the string Equation (3.24) used there

$$\frac{\partial^2 y}{\partial x^2} = \frac{\lambda}{T}\frac{\partial^2 y}{\partial x^2} \tag{8.1}$$

in terms of propagation velocity c and angular frequency ω to get

$$\frac{\partial^2 y}{\partial x^2} = \frac{\omega^2}{c^2}\frac{\partial^2 y}{\partial x^2} \tag{8.2}$$

Following Montgomery et al. (2003), if we make a small position-dependent perturbation $\delta c(x)$ to the propagation velocity, then we can use a variational principle to estimate the resulting relative frequency shift,

$$\frac{\delta\omega_n}{\omega_n} = \frac{2}{L}\int_0^L dx\left[\frac{\delta c(x)}{c}\right]\sin^2(k_n x) \tag{8.3}$$

[1] Or, in one or two cases, have special interest to me. I admit it.

doi:10.1088/2514-3433/ae03a0ch8

which more generally we can write as

$$\frac{\delta\omega_n}{\omega_n} = \int_0^L dx \left[\frac{\delta c(x)}{c}\right] K_{n,c} \tag{8.4}$$

Here $K_{n,c}$ is the *kernel* for the nth eigenfunction; essentially, the unperturbed solution. It's clear that differing perturbations to c will have influence some modes more than others, and that the overlap between the spatial contributions from the perturbation and the kernel determine the extent of that influence. Eigenfunctions are affected as well, with sharp changes in the propagation velocity leading to "kinks" in the eigenfunctions (see Section 8.4 below).

For the case of a stellar model, the equivalent expression is

$$\frac{\delta\omega_i}{\omega_i} = \int_0^R K^i_{c^2,\rho}(r)\left(\frac{\delta c}{c}\right)^2 dr + \int_0^R K^i_{\rho,c^2}(r)\left(\frac{\delta\rho}{\rho}\right) dr \tag{8.5}$$

Here the two functions K^i represent the changes in frequency due to changes in sound speed c and density ρ,[2] and are calculated based on a reference model.

8.1.2 Applications to Stars

So far in this volume, approaches to converting oscillation frequencies to actual physical characteristics of the star (mass, radius, age, structure) have broadly fallen into two categories. The first is to use scaling relationships based on the average large frequency separation $\Delta\nu$ or the frequency of maximum power $\nu_{\rm max}$ to infer fundamental stellar characteristics. This approach has the virtue of simplicity and limited computational demands, but its accuracy is limited by both the calibration of the relationship used and the degree to which the adopted scaling relationship is truly universal. Alternatively, one can adopt a forward approach, calculating a grid of models with a stellar interiors code such as MESA, and comparing the model frequencies to the observed frequencies. Non-asteroseismic constraints, such as spectroscopy, or seismic scaling relations can be used to limit the parameter space and thus the computational requirements for the grid, and a least-squares or similar technique can be used to find the best-fit models. Interpolation across the model grid then gives the stellar parameters. Alternatively, the grid can be calculated more on-the-fly and a genetic algorithm used to find the best fit. Either approach makes use of the seismic information embedded in the individual frequencies, a more complete set than the approximation inherent in the scaling relations, at the cost of a significant increase in computational expense. In addition, while using the scaling relations implies reliance on the validity of the relationship and its calibration, model grids are dependent on the specific stellar models used.

An alternative approach, used for many years in helioseismology, is to perform a structural inversion, to use the frequency data to directly infer the structure of the

[2] Other kernel pairs are also possible.

star. In general, the observed frequencies depend on the internal density, sound speed, rotation rate, magnetic field, and other structural properties, so

$$\nu_{nl} = f[\rho(r), c(r), \ldots] \tag{8.6}$$

The dependence here is nonlinear, which greatly complicates the inversion. We work around this problem by making use of a reference model, whose parameters $\rho_0(r)$, $c_0(r)$, ... are close to the star of interest, and correspond to the oscillation frequencies ν_0 of the reference model. In that case, we are searching for the radial corrections

$$\begin{aligned} \delta_r \rho(r) &= \rho(r) - \rho_0(r) \\ \delta_r c(r) &= c(r) - c_0(r) \\ &\cdots \end{aligned} \tag{8.7}$$

and similarly for the nonradial dependencies. If the differences are small, we can linearize the problem, so that

$$\frac{\delta \nu_{nl}}{\nu_{nl}} = \int_0^R K^{nl}_{\rho,\, c,\, \ldots}(r) \frac{\delta c(r)}{c(r)} dr + K^{nl}_{\rho,\, c,\, \ldots}(r) \frac{\delta \rho(r)}{\rho(r)} dr \tag{8.8}$$

Here the kernel functions K^{nl} characterize the changes in frequency due to (small) changes in the physical parameters ρ, c, ... We calculate those functions based on the reference model. Formally, we can also include terms for the surface correction and observational errors. Figure 8.1 illustrates the appearance of some structure kernels for the subgiant HR 7322 observed by Kepler.

The equations contain the data (in terms of the frequencies) and the kernels (calculated from the chosen reference model), so what's missing are the relative changes of the physical parameters $\frac{\delta c(r)}{c(r)}$, $\frac{\delta \rho(r)}{\rho(r)}$, etc. In theory this is a linear system of equations, so we can solve for those functions, but in reality the problem is ill-posed because in the stellar case typically we only have at best tens of modes available to constrain the functions $\rho(r)$, $c(r)$, ..., which effectively limits our resolution of the structural parameters as functions of radius.

Practical solution techniques fall into two families. The first are least-squares methods, which as usual seek to minimize a parameter

$$\chi^2 = \sum \left(\frac{\frac{\delta \nu_{nl}}{\nu_{nl}} - \Delta \nu_{nl}}{\sigma_{nl}} \right)^2 \tag{8.9}$$

However, in practice the solution is poorly constrained so the solution displays unphysical oscillatory behavior, which is addressed by *regularizing* the χ^2 parameter such that

$$\chi^2_{reg} = \chi^2 + \lambda R \tag{8.10}$$

where R is a penalty function, typically related to the second derivative of the functions.

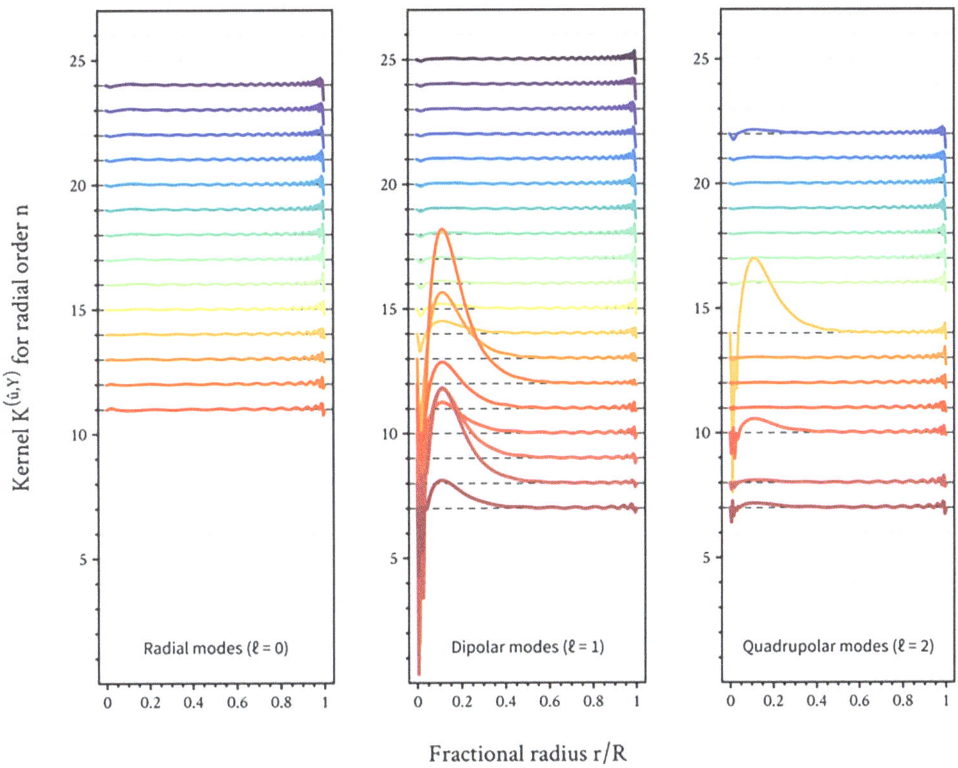

Figure 8.1. Structure kernels showing the sensitivity of each mode's frequency to a perturbation in sound speed. The quadrupolar $l-2$ modes with radial orders 9 and 15 were not detected, probably because of their high mode inertias. The zero-point of each kernel is offset by its radial order, and each kernel is normalized by the same fixed value and colored for visibility. Reprinted with permission from Bellinger et al. (2021).

The second family of approaches is based on optimally localized averages (OLA). Here we calculate inversion coefficients c_i chosen so that most of the contribution to the sum

$$\mathscr{K}(r_0, r) = \sum_i c_i(r_0) K^i_{c_s^2,\,\rho}(r) \tag{8.11}$$

comes from $r \approx r_0$. Here $\mathscr{K}(r_o, r)$ is the *averaging kernel* at $r = r_0$ and if it is normalized so that

$$\int \mathscr{K}\, dr = 1 \tag{8.12}$$

then we will have

$$\left(\frac{\delta c_s^2}{c_2^2}\right)_{r_0} = \sum_i c_i(r_0) \frac{\delta\omega_i}{\omega_i} \tag{8.13}$$

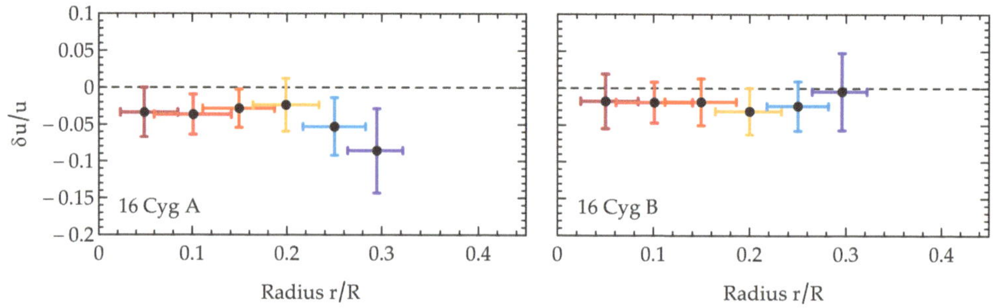

Figure 8.2. Structural inversions for the internal squared isothermal sound-speed profile u of 16 Cyg A (left) and 16 Cyg B (right). In both cases, core sound speeds are greater than those from the reference models. Reprinted with permission from Bellinger et al. (2017).

which represents an average of the relative sound speed differences at r_0 relative to the stellar model. The complexity here comes from constructing an averaging kernel at the r_0 that we are interested in! This is only possible if there's sufficient information about the sound speed at that depth, which means we need "enough" modes with lower turning points close to r_0. A number of techniques (see Bellinger 2018; Bétrisey 2024, and references therein) have been developed to construct the averaging kernels.

Only recently has the number of solar-like oscillators with sufficient modes to make the inversion process meaningful grown significantly, but applications of the technique are increasing in popularity. One recent example is shown in Figure 8.2.

8.2 Rotation

The fundamental description of stellar oscillations in terms of spherical harmonics assumes spherical symmetry. Any deviation from that case will affect the observed frequencies, and rotation is certainly such a deviation. Intuitively, considering a uniformly rotating star such that $\phi = \phi_0 + \Omega t$, one can compare the behavior of a wave propagating in the prograde direction around the stellar equator to one in the retrograde direction, and realize that in the observer's reference frame the two have different periods.

We can of course write the linearized equations of stellar oscillations as displaying a complex periodicity $\exp(-i\omega t)$. If in turn we explicitly write the frequency in real and complex components, so that $\omega = \omega_{\rm real} + i\omega_{\rm complex}$, the dependence becomes

$$\cos(m\phi - \omega_{\rm real} t + \delta_0)\exp(i\omega_{\rm complex} t) \tag{8.14}$$

Here δ_0 is the phase at time $t = 0$. For $m = 0$ we recover a simple standing wave, while for nonzero m the wave travels around the equator with a period $2\pi/\omega_{\rm real}$. In this description, the imaginary component of the frequency, which leads to the $\exp(\omega_{\rm complex} t)$ term in Equation (8.14), describes the growth rate of the wave; for simplicity, we here take this term to be unity so that the wave amplitude remains constant.

Applying this description to rotation requires recognition that the constantly rotating star has different coordinate descriptions in the reference frames of the star and the observer. If the stellar frame is described by (r, θ, ϕ) and we define the rotation as occurring in the ϕ direction, then $\phi \rightarrow \phi - \Omega t$ and the outside observer sees

$$\cos(m\phi - \omega) \rightarrow \cos(m[\phi - \Omega t] - \omega) = \cos(m\phi - [\omega + m\Omega]) \tag{8.15}$$

where we've taken $\delta_0 = 0$ for clarity. Effectively, the frequencies are split uniformly by the rotation rate Ω scaled by the azimuthal order m, as described above in Equation (6.31):

$$\omega_{nlm} = \omega_{nl} + m(1 - C_{nl})\Omega \tag{8.16}$$

where Ω is the stellar angular rotation rate and $0 \leqslant C_{nl} \leqslant 1$ is the *Ledoux constant*.

This simple description neglects a number of factors that can become important. Rotation changes the *shape* of the star through a centripetal force contribution to the force balance, and the Coriolis force also contributes to differences between the descriptions of the motion by observers in the rotating and non-rotating frames. Perhaps most importantly, the rigid-body constant rotation approximation is certainly incorrect, as in general we would anticipate that rotation varies significantly as a function of radius, and locally as a function of latitude as well, though observing the latter effect in stars other than the Sun is likely to be challenging, and that sufficiently rapid rotation will change the structure of the star.

A somewhat more sophisticated approach involves using the Traditional Approximation of Rotation (TAR; Mathis & Prat 2019). In a rotating star, the buoyancy force is supplemented by the Coriolis force as a restoring force for oscillations, leading to the nomenclature of *gravito-inertial modes*. In the TAR approximation, the Coriolis force in the radial direction is neglected, which allows treatment of the gravito-inertial waves as essentially horizontal. This in turn allows separability of the equations of motion in the vertical and horizontal direction just as we have done for non-rotating stars. Formally, the TAR requires that the buoyancy force be stronger than the Coriolis force ($2\Omega < N$, where Ω is the rotational and N the buoyancy frequency), that N be larger than the mode frequency in the rotating frame, and that the star can be treated as spherical and uniformly rotating. In practice, this limits application to low-frequency gravito-inertial modes in stars that aren't rotating very fast, but nonetheless provides significant insight. Figure 8.3 illustrates the effects of rotation on g-mode period spacing patterns for a 12 solar mass star (Bowman 2020).

8.3 Magnetic Fields

There are (so far) two fundamental ways with which to detect internal magnetic fields using asteroseismology. The first was mentioned earlier, in Section 6.4, and involves asymmetries in the line shifts observed in rotationally split multiplets. This was originally predicted for $l = 1$ g-modes in SPB stars by Hasan et al. (2005), and more sophisticated models by Loi (2020) and Bugnet et al. (2021) for g-modes and

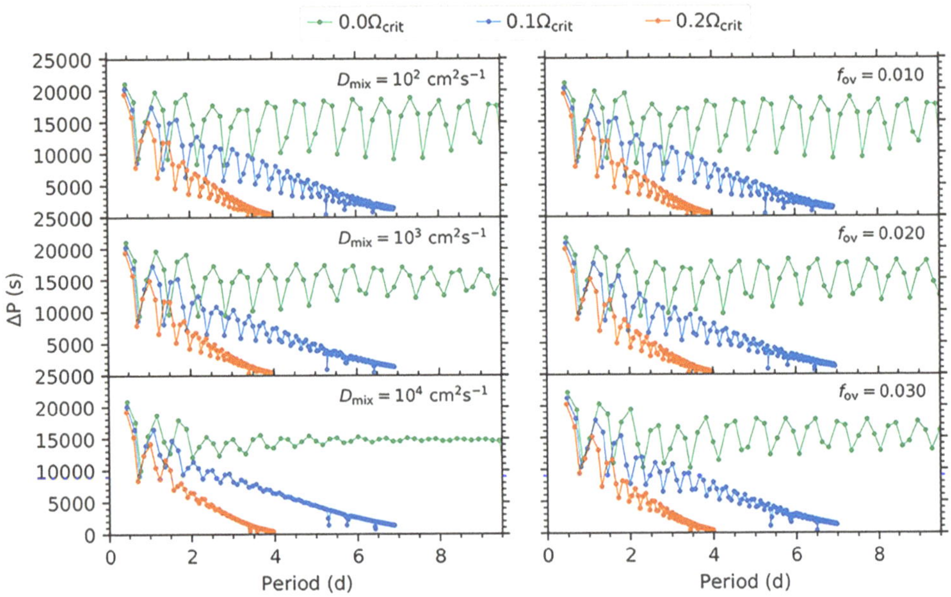

Figure 8.3. Predicted g-mode period spacing patterns for prograde modes of a 12 solar mass star midway through its main sequence lifetime, for rotation rates in terms of the critical (breakup) rotational velocity. Plots are labeled for different envelope mixing parameters ($D_{\rm mix}$) and convective overshoot ($f_{\rm ov}$), the latter expressed in terms of the local pressure scale height. Model calculations were done using MESA (Paxton et al. 2019) using the TAR, and frequencies calculated using GYRE (Townsend & Teitler 2013). Reprinted with permission from Bowman (2020).

mixed modes, respectively, suggested that fields in the range of 100–1000 kG should be detectable.

Li et al. (2022) detected asymmetric splittings in mixed modes of red giant stars using Kepler (Figure 8.4, and the magnitude of the shifts measured by Li et al. (2022) for red giants and Lecoanet et al. (2022) for a main sequence B star were both in the range of a few $\times 10^5$ G, in accordance with expectations (see Figure 8.5).

A different way to detect stellar magnetic fields with asteroseismology was suggested by Fuller et al. (2015) and Stello et al. (2016), who found that stars with strong internal magnetic fields show suppressed dipole ($l = 1$) modes, while retaining radial modes of normal amplitude. If the core has a strong B field, then the oscillation has a nonzero transmission coefficient through the core, and loses energy during the time it takes to cross. The net result is significantly increased damping, which both lowers the amplitude and increases the linewidth of the mode (by decreasing its lifetime).

Stello et al. (2016) calculated that visibility of $l = 1$ dipole modes should be suppressed by roughly a factor of 3 for red giant stars with $\nu_{\rm max} = 70\ \mu Hz$ and that factor should increase rapidly as $\nu_{\rm max}$ decreases. Figure 8.6 confirms that roughly 20% of their sample showed the effect, and are presumed to have strong core magnetic fields. More recent work (Lin et al. 2025) showed that effect extended to subgiants as well, though less commonly.

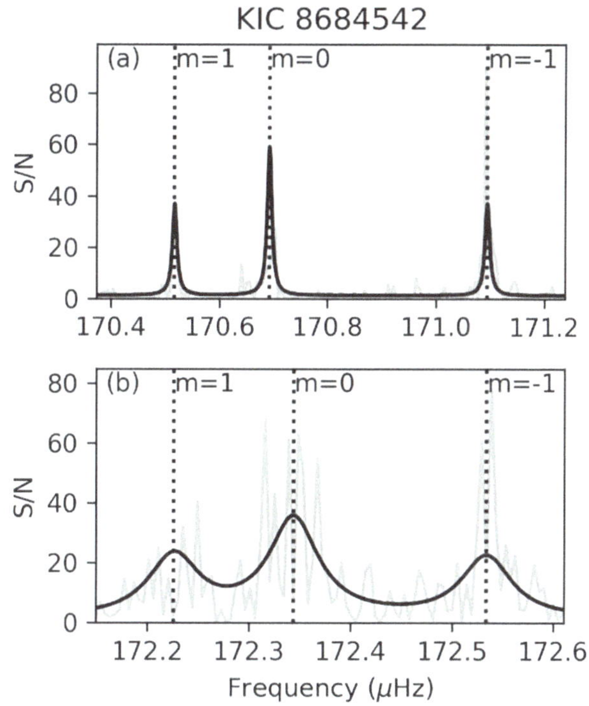

Figure 8.4. Asymmetric splittings of two mixed modes in the red giant star KIC 8684542, observed with Kepler. The upper panel shows splittings in g-dominated modes, and the lower in p-dominated. Gray lines show observations and overlying black lines the best-fitting model spectra, with the dotted lines indicating the frequency centers. Reprinted with permission from Li et al. (2022).

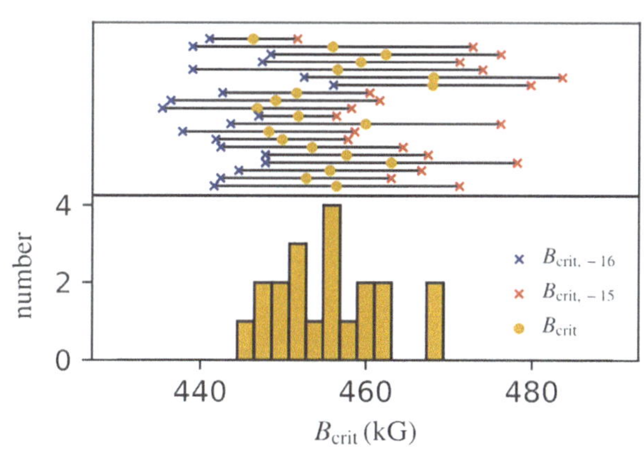

Figure 8.5. Top: Inferred magnetic field strengths or the 20 best-fitting asteroseismic models for the B star HD 43317. The blue (red) crosses show the field strength for the $l = 2, n = -16$ ($n = -15$) modes, while the yellow circles are their mean, taken as an estimate of the field strength, inferred to be $B_r = 456 \pm 6\text{k}G$. Reprinted with permission from Lecoanet et al. (2022).

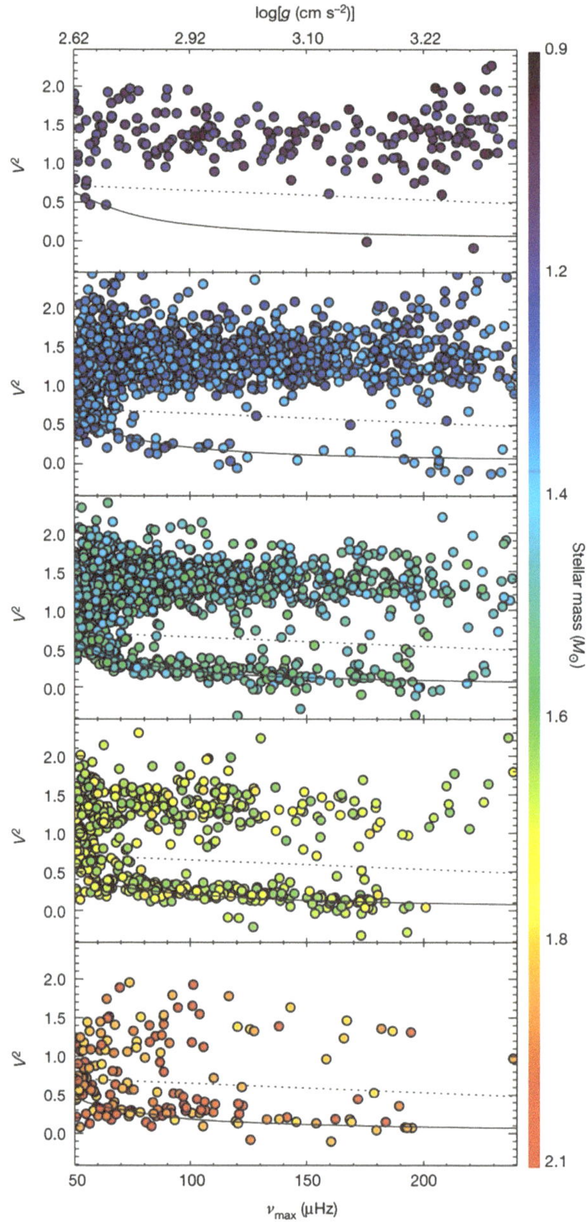

Figure 8.6. Visibility against $\nu_{\max}$ for a sample of Kepler giants. Stars in the diagram evolve from right to left, from roughly the start of the RBG to the luminosity bump, and color-coding indicates mass, with an uncertainty of about 10%. The solid black line shows the theoretical predicted dipole-mode visibility suppression for 1.1, 1.3, 1.5, 1.7, and 1.9 $M_\odot$ models and a radial mode lifetime of 20 days. The dashed line is a nominal separation between normal and suppressed stars. Reprinted with permission from Stello et al. (2016).

8.4 Glitches

Glitches occur when the radial sound speed profile of the star contains abrupt changes, typically at the radiative/convective boundary, or at a partial ionization zone, since the sound speed c follows

$$c^2 = \frac{\gamma P}{\rho} \tag{8.17}$$

and of course the value of γ changes in a partial ionization zone. The effect, as described by Gough & Thompson (1988) and Gough (1990), is to introduce an oscillatory component into the *eigenfrequencies*, such that the frequencies themselves oscillate,

$$\delta\nu_{\rm glitch} \propto \sin(4\pi\tau_{\rm glitch}\nu_{nl} + \phi) \tag{8.18}$$

Here $\tau_{\rm glitch}$ is the acoustic depth of the glitch,

$$\tau_{\rm glitch} = \int_{r_{\rm glitch}}^{R_S} \frac{dr}{c} = \frac{1}{2\Delta\nu} \tag{8.19}$$

R_S is the *acoustic* radius of the star, which in practice lies close to the physical radius.

The effect is small, so the highest quality data is necessary, but analysis of glitches allows measurement of the depth of the convection zone as well as the location of the He II ionization zone. In the latter case, the amplitude of the glitch is proportional to the helium abundance, so can provide an independent measurement of Y in addition to that derived from spectroscopy or interiors model fits.

An example of the predicted behavior based on model calculations is shown in Figure 8.7, while Figure 8.8 shows observations of the effect made using Kepler. Both trace glitches by examining second differences in frequencies, such as

$$\Delta_2\nu_{n,l} = \nu_{n-1,l} - 2\nu_{n,l} + \nu_{n+1,l} \tag{8.20}$$

8.5 Mode Identification

In order to go beyond simple applications of scaling relations to true modeling, it it critical to understand which modes nl are actually responsible for the measured frequencies that are observed, a problem we call *mode identification*. In solar-like oscillators, the regular pattern in the power spectrum is clearly visible in both the power spectrum and the echelle diagram, making mode identification relatively straightforward. However, as we move up the main sequence, things become more complicated. As Figure 6.8 (central panel) shows, F stars have relatively short mode lifetimes, which manifests as a broadening of the peaks in the power spectrum and the width of the ridges in the echelle diagram, leading to confusion between neighboring ridges and uncertainty regarding whether a given peak is $l = 0$ or $l = 2$. The situation gets even worse for intermediate-mass stars; here, modes are not stochastically excited, and the broad envelope of power that is seen in cooler stars is absent, so there are generally no clues as to the mode identification of the observed

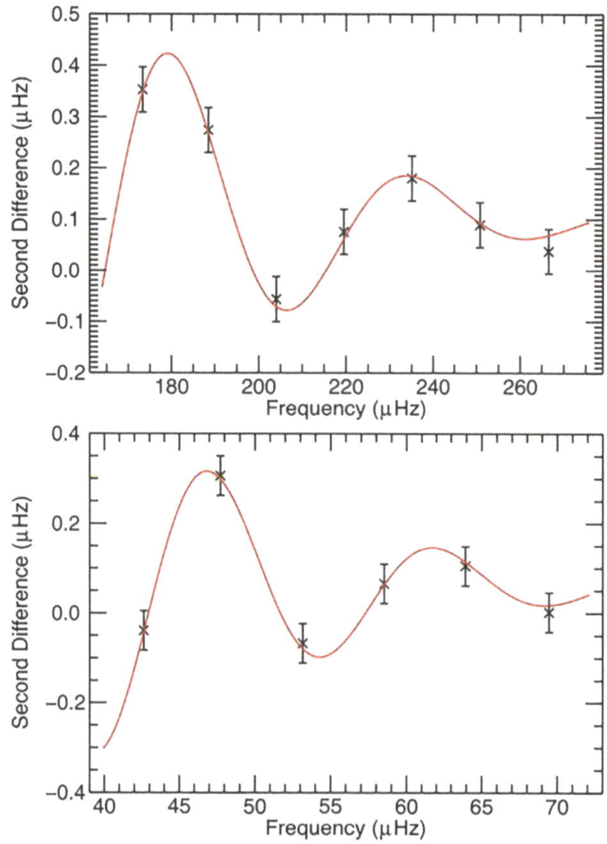

Figure 8.7. Calculated examples of the $l = 0$ second differences (black crosses) predicted by two different red giant models, as a function of frequency. Reprinted with permission from Broomhall et al. (2014).

peaks. The existence of rotational splittings and mixed modes can also complicate the power spectrum sufficiently to make mode identification by inspection impossible.

8.5.1 Surface Phase

The asymptotic relation for p-modes is

$$\nu_{nl} = \Delta\nu\left(n + \frac{l}{2} + \alpha + \frac{1}{4} - l(l+1)D_0\right) \tag{8.21}$$

Observationally, this is typically written as

$$\nu_{nl} = \Delta\nu\left(n + \frac{l}{2} + \varepsilon - \delta\nu_{0l}\right) \tag{8.22}$$

where ε is an offset that can be interpreted as a phase factor determined by conditions near the upper and lower turning points of the mode (Gough 1993), in

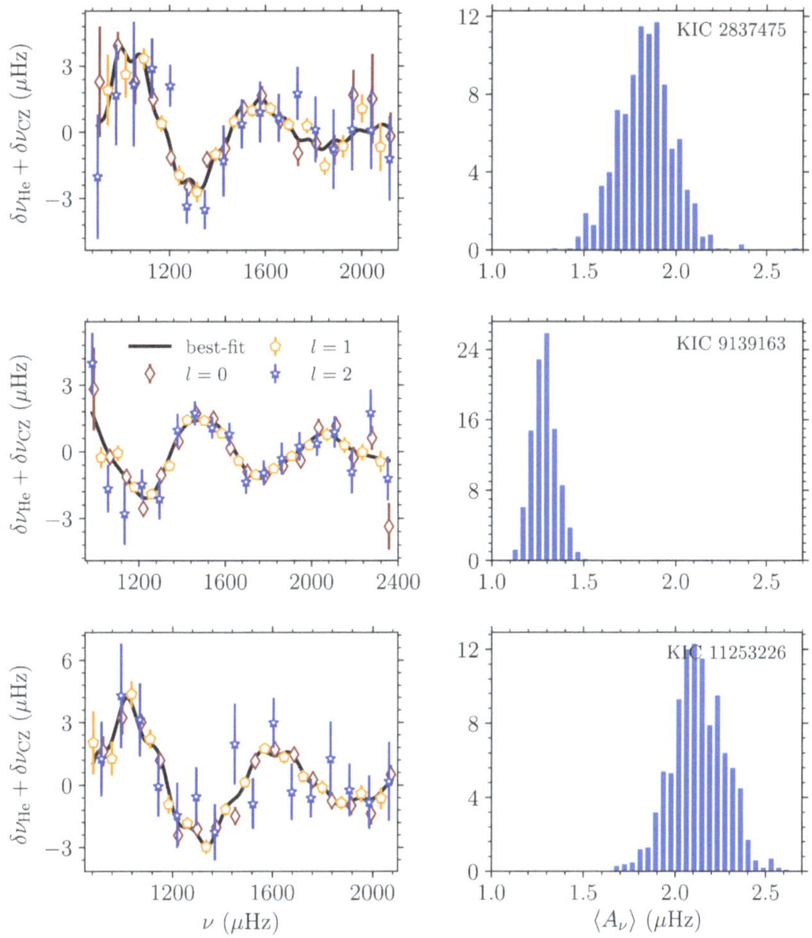

Figure 8.8. The smooth overall fit to the observed oscillation frequencies for three stars observed with Kepler has been subtracted to clearly show the glitch signatures). The left-hand panels show different observed modes of harmonic degrees $l = 0$, 1, and 2, while the curve represents the best fit to them. The histograms in the right-hand panels show the distribution of average amplitude of the helium signature obtained using a Monte Carlo simulation. Reprinted with permission from Verma & Silva Aguirre (2019).

particular the stellar atmosphere, which is typically included in asteroseismic models in a relatively primitive way. White et al. (2012) (see also White et al. 2011a, 2011b) pointed out that this implied a relationship between ε and $T_{\rm e}$. White et al. (2012) calibrated this relationship using Kepler stars with unambiguous identifications, and were then able to successfully use the resulting relationship to uniquely identify modes in stars for which two different mode identifications were both plausible.[3]

[3] It is also worth noting that Ong & Basu (2019) point out that properly the phase factor ε should be considered as a function of ν_{nl}, and that $\varepsilon(\nu_{nl})$ encodes additional information, particularly about the stellar evolutionary state; in particular, it can be used to distinguish between stars on the ascending portion of the RGB and those in the red clump.

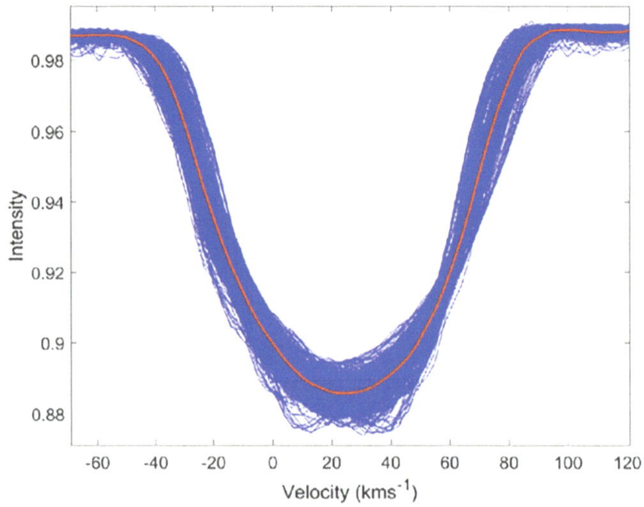

Figure 8.9. Line profiles of 613 observations of γ Doradus, with the mean profile overplotted in red. This star shows large pulsational variation from observation to observation. Reprinted with permission from Brunsden et al. (2018).

8.5.2 Spectroscopic Moments

The velocity pattern in the photosphere of an oscillating star varies depending on the spherical harmonic *nl* of the mode, and Balona (1986) pointed out that this signature could be detected from the shapes of spectral lines using high-resolution spectroscopy. Figure 8.9 shows the variability possible in a pulsating star. Aerts & Waelkens (1993) and Aerts (1996) extended the procedure to fully develop the *moment method.*

In this approach, the first three moments of the line profile are calculated for each possible mode, and compared with observations, along with phase parameters for each fit (Figure 8.10). An alternate approach based on the same underlying physics is the pixel-by-pixel method (Mantegazza 2000). This is a line-profile variation method that looks at the movement of each individual pixel in a spectral absorption line and treats them as a time series, with each pixel used to create an average Fourier spectrum of the frequencies present in the line profile.

8.5.3 Amplitude Ratios

As the shape of the outer layers of the star changes, the effective temperature T_e and surface gravity g also change locally, leading to changes in the integrated flux over the course of a pulsational cycle (Dziembowski 1977; Stamford & Watson 1981; Watson 1988; Garrido et al. 1990; Garrido 2000; Dupret et al. 2003). The flux as a function of time then is

$$F_\lambda + \delta F_\lambda(\theta, \phi, t) = F\,[T_{e,0} + \delta T_e(\theta, \phi, t), g_0 + \delta g_e(\theta, \phi, t)] \tag{8.23}$$

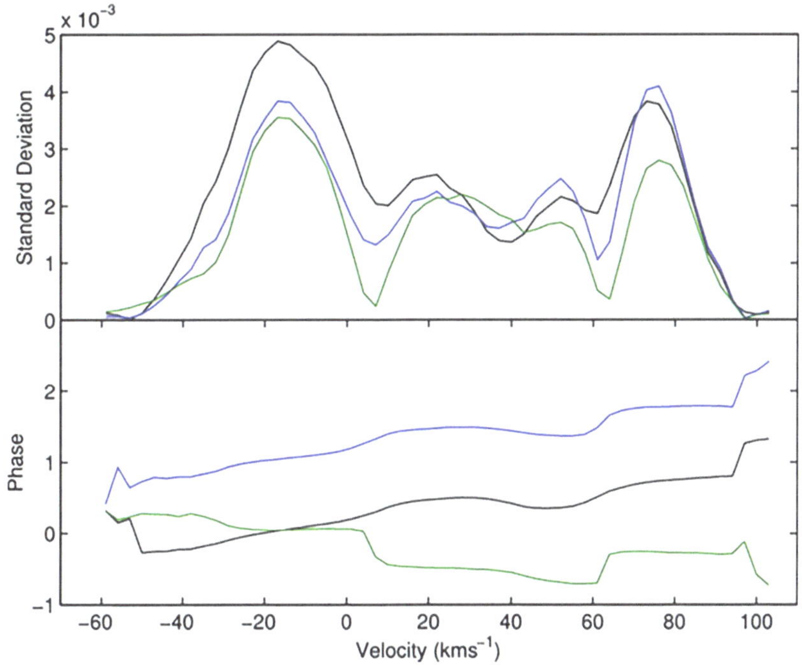

Figure 8.10. Comparison of the standard deviation and phase profiles for three frequencies, $f_3 = 0.3167\ \mathrm{d}^{-1}$ (black), $f_3 + 1\ \mathrm{d}^{-1}$ (blue) and $1.32098\ \mathrm{d}^{-1}$ (green), illustrating the range of variation possible. Reprinted with permission from Brunsden et al. (2018).

In the linear regime where changes are small, we can perturb this expression to derive

$$\frac{\delta F_\lambda}{F_\lambda} = \frac{\partial \log F_\lambda}{\partial \log T_\mathrm{e}} \frac{\delta T_\mathrm{e}}{T_\mathrm{e}} + \frac{\partial \log F_\lambda}{\partial \log g_\mathrm{e}} \frac{\delta g_\mathrm{e}}{g_\mathrm{e}} \tag{8.24}$$

We can also write the fractional radial displacement of the photosphere in relative terms as

$$\varepsilon = \frac{\xi_\mathrm{r}(\theta, \phi, t)}{R P_l^m(\cos\theta)\cos(\omega t + m\phi)} \tag{8.25}$$

making use of the $P_l^m(\cos\theta)$ to associate the displacement with the appropriate spherical harmonic for each mode. Limb darkening can be included using a similar formalism, so that we have

$$\begin{aligned} \frac{\delta T_\mathrm{e}}{T_\mathrm{e}}(\theta, \phi, t) &= f_T\, \varepsilon P_l^m(\cos\theta)\cos(\omega t + m\phi + \psi_T) \\ \frac{\delta g_\mathrm{e}}{g_\mathrm{e}}(\theta, \phi, t) &= -f_g\, \varepsilon P_l^m(\cos\theta)\cos(\omega t + m\phi) \end{aligned} \tag{8.26}$$

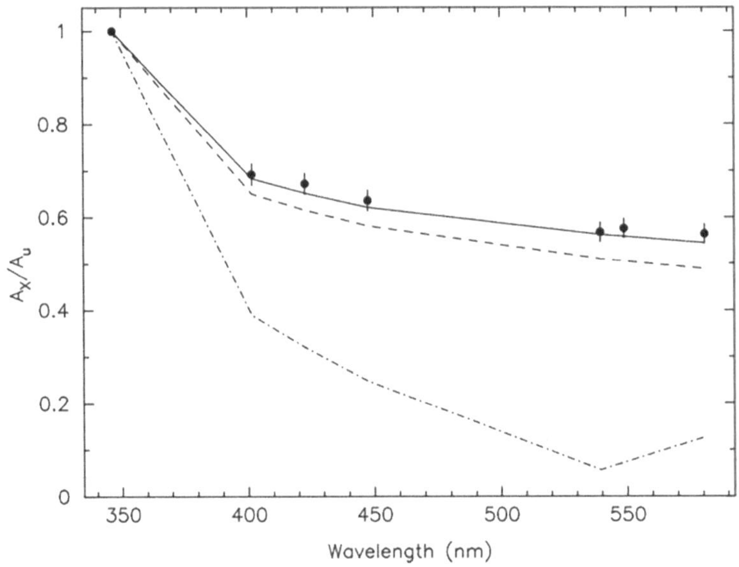

Figure 8.11. Predicted and observed Geneva photometry amplitude ratios for the dominant mode of the SPB star HD 74560. The points with error bars represent the observations, while the lines correspond to theoretical predictions for different angular degrees: $l = 1$ (solid), $l = 2$ (dashed), and $l = 3$ (dot-dashed). Reprinted with permission from Dupret et al. (2003).

where the minus sign in the second equation accounts for the fact that δg_e is oppositely directed from the radial displacement $\delta\xi_r$, and the ψ_T allows changes in T_e and g_e to be out of phase with one another.

We don't actually know the relative displacement ε, so we use flux ratios (or, equivalently, magnitude differences), and a model atmosphere to calculate the partial derivatives in Equation 8.24. Figure 8.11 shows an application for the SPB star HD 74560.

8.5.4 Polarization

Polarization can be induced by a number of different astrophysical processes, including magnetic fields, rapid rotation, reflection, and scattering. As far back as 1946, Chandrasekhar (1946) predicted that electron scattering in the atmospheres of hot stars would result in linear polarization, increasing radially from disk center to the stellar limb. In a static spherical star viewed by a distant observer as a point source, the total polarization would average to zero. However, departures from sphericity, caused for example by rapid stellar rotation, can lead to measurable net polarimetric signals. More importantly for our case, oscillations change the shape of the star and can also lead to polarimetric variability.

The *type* of pulsation matters here, though. Clearly, radial pulsations will not change the net zero linear polarization expected from a spherically symmetric source. Even $l = 1$ nonradial modes will not do that, because polarization is a *pseudovector* rather than a true vector, but higher degree modes imprint a signal on

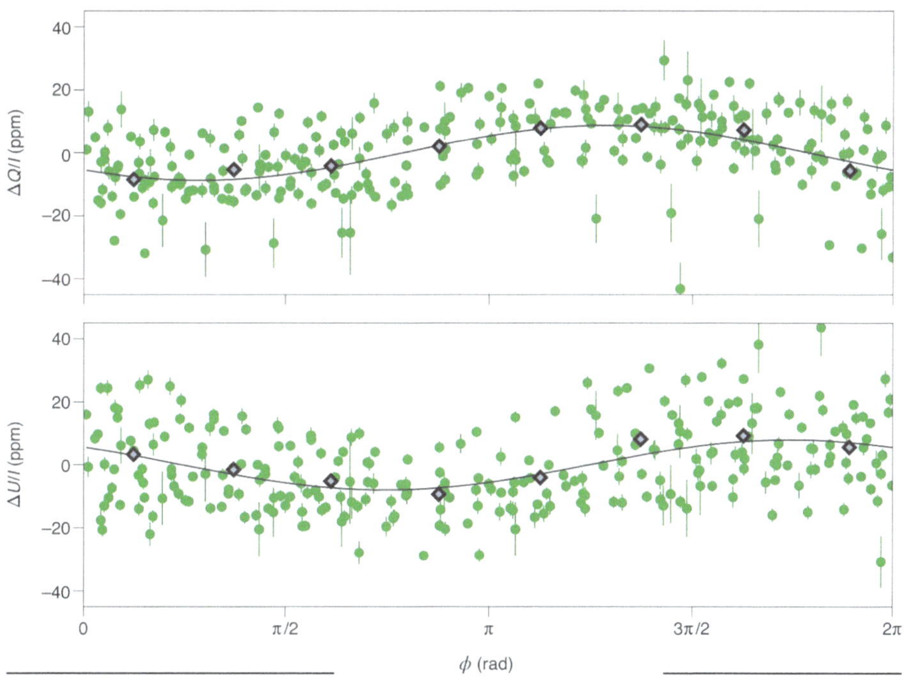

Figure 8.12. The change in polarization for the β Cep star β Cru, presented as normalized Stokes vectors Q/I (top) and U/I (bottom), as a function of phase, ϕ. The data is phase-folded to correspond to photometric frequency $f_2 = 5.964\text{d}^{-1}$. Data points with 1 σ errors are shown in green, while the sinusoidal fit is in black; the gray points are data binned to $\pi/4$ rad. Reprinted with permission from Cotton et al. (2022).

the polarization position angle. Figure 8.12 illustrates a polarimetric detection of an $l = 3$ oscillation mode in the β Cep star β Cru (Cotton et al. 2022). Note that part-per-million precision in polarimetry is necessary to detect the variability. Combining these data with space-based photometry and spectroscopy when available, allows the universe of possible modes for each frequency to be dramatically reduced and the grid modeling required to be rendered a tractable problem.

8.6 Machine Learning

Over the last decade, machine learning (ML) has made aggressive entry into science in general and astronomy in particular, aided by the fact that our field has large data sets and lots of classification and discovery to be done. To date, within asteroseismology, ML has been applied in a range of areas, including

- Discovery: TESS will observe millions of stars by the end of its mission, and the upcoming PLATO mission[4] will continue to add even more. Building a full-sky catalog of variable stars from both missions and determining which are p- and g-mode oscillators will be a significant task for ML (Audenaert et al. 2024). Even in cases where human intervention is needed, Davies et al.

[4] https://www.esa.int/Science_Exploration/Space_Science/Plato_factsheet

(2016) has demonstrated that a Bayesian unsupervised machine learning approach can be used for improved quality assurance in determining oscillation frequencies.

- Classification: The number of new light curves already vastly exceeds the ability of human beings[5] to sift through and classify,[6] and this is a tailor-made role for ML. Hon et al. (2018) have demonstrated that a convolutional neural net classifier can be trained to classify stellar evolutionary states based even on lower frequency resolution, lower signal-to-noise frequency spectra. Wu et al. (2019) and Zinn et al. (2020) have demonstrated that ML can differentiate between stars in the red clump and those on the RGB, and Le Saux et al. (2019), Barbara et al. (2022), and Audenaert et al. (2021) have successfully used ML tools for broader classification of oscillating stars.
- Automated measurement: ML tools have been trained to measure frequency and period spacings in red giants (Dhanpal et al. 2022), oscillation peak patterns that are useful to infer mean stellar densities and differential rotation (Mirouh et al. 2021), and even perform mode identification in intermediate and high-mass stars (Mirouh 2022).
- Physical inference: Grid-based modeling is computationally extremely expensive, and inversions are still only possible for a small fraction of targets. More than a decade ago, Metcalfe et al. (2014) used a parallel genetic algorithm (Metcalfe & Charbonneau 2003) to perform a more directed search of parameter space, and modern ML tools can certainly improve on this. So far, Hon et al. (2024) have used a flow-based generative approach to emulate grids of stellar evolutionary models to relatively rapidly determine improved mass and radius estimates for over 15,000 red giants, and Bellinger et al. (2016) have used it to rapidly estimate fundamental stellar parameters from combined classical and asteroseismic observations; see also Li et al. (2022) and Panda et al. (2024).

References

Aerts, C. 1996, A&A, 314, 115

Aerts, C., & Waelkens, C. 1993, A&A, 273, 135

Audenaert, J., Tkachenko, A., Skarka, M., Eschen, Y. N. E., & Muthukrishna, D. 2024, 8th TESS/15th Kepler Asteroseismic Science Consortium Workshop (Lisbon: Instituto de Astrofísica e Ciências do Espaço) 47

Audenaert, J., Kuszlewicz, J. S., Handberg, R., et al. 2021, AJ, 162, 209

Balona, L. A. 1986, MNRAS, 219, 111

Barbara, N. H., Bedding, T. R., Fulcher, B. D., Murphy, S. J., & Van Reeth, T. 2022, MNRAS, 514, 2793

Bellinger, E. P. 2018, PhD Thesis, Max-Planck-Institute for Solar System Research, Lindau

Bellinger, E. P., Angelou, G. C., Hekker, S., et al. 2016, ApJ, 830, 31

Bellinger, E. P., Basu, S., Hekker, S., & Ball, W. H. 2017, ApJ, 851, 80

[5] There aren't enough students, and it's not nice to make them do it, either!

[6] Of course, the most interesting targets may well be those that resist ready classification.

Bellinger, E. P., Basu, S., Hekker, S., Christensen-Dalsgaard, J., & Ball, W. H. 2021, ApJ, 915, 100
Bétrisey, J. 2024, PhD Thesis, University of Geneva, Switzerland
Bowman, D. M. 2020, FrASS, 7, 70
Broomhall, A. M., Miglio, A., Montalbán, J., et al. 2014, MNRAS, 440, 1828
Brunsden, E., Pollard, K. R., Wright, D. J., De Cat, P., & Cottrell, P. L. 2018, MNRAS, 475, 3813
Bugnet, L., Prat, V., Mathis, S., et al. 2021, A&A, 650, A53
Chandrasekhar, S. 1946, ApJ, 103, 351
Cotton, D. V., Buzasi, D. L., Aerts, C., et al. 2022, NatAs, 6, 154
Davies, G. R., Silva Aguirre, V., Bedding, T. R., et al. 2016, MNRAS, 456, 2183
Dhanpal, S., Benomar, O., Hanasoge, S., et al. 2022, ApJ, 928, 188
Dupret, M. A., De Ridder, J., De Cat, P., et al. 2003, A&A, 398, 677
Dziembowski, W. 1977, AcA, 27, 203
Fuller, J., Cantiello, M., Stello, D., Garcia, R. A., & Bildsten, L. 2015, Sci, 350, 423
Garrido, R. 2000, Delta Scuti and Related Stars, Reference Handbook and Proc. of the 6th Vienna Workshop in Astrophysics ed. M. Breger, & M. Montgomery (San Francisco, CA: ASP) 67
Garrido, R., Garcia-Lobo, E., & Rodriguez, E. 1990, A&A, 234, 262
Gough, D. O. 1990, Progress of Seismology of the Sun and Stars (Vol. 367, ed. Y. Osaki, & H. Shibahashi; Berlin: Springer-Verlag) 283
Gough, D. O. 1993, 47th Session de l'Ecole d'Eté de Physique Théorique: Astrophysical fluid dynamics - Les Houches 1987, ed. J. P. Zahn, & J. Zinn-Justin (Amsterdam: Elsevier) 399
Gough, D. O., & Thompson, M. J. 1988, IAU Symp., Vol. 123, Advances in Helio- and Asteroseismology, ed. J. Christensen-Dalsgaard, & S. Frandsen (Dordrecht: Reidel Publishing Co.) 155
Hasan, S. S., Zahn, J. P., & Christensen-Dalsgaard, J. 2005, A&A, 444, L29
Hon, M., Li, Y., & Ong, J. 2024, ApJ, 973, 154
Hon, M., Stello, D., & Zinn, J. C. 2018, ApJ, 859, 64
Le Saux, A., Bugnet, L., Mathur, S., Breton, S. N., & García, R. A. 2019, SF2A-2019: Proc. of the Annual Meeting of the French Society of Astronomy and Astrophysics, ed. P. Di Matteo, et al. (Paris: SF2A)
Lecoanet, D., Bowman, D. M., & Van Reeth, T. 2022, MNRAS, 512, L16
Li, G., Deheuvels, S., Ballot, J., & Lignières, F. 2022, Natur, 610, 43
Lin, S., Li, T., Mao, S., & Fuller, J. 2025, ApJ, 980, 217
Loi, S. T. 2020, MNRAS, 496, 3829
Mantegazza, L. 2000, ASP Conf. Ser., Vol. 210, Delta Scuti and Related Stars ed. M. Breger, & M. Montgomery (San Francisco, CA: ASP) 138
Mathis, S., & Prat, V. 2019, A&A, 631, A26
Metcalfe, T. S., & Charbonneau, P. 2003, JCoPh, 185, 176
Metcalfe, T. S., Creevey, O. L., Doğan, G., et al. 2014, ApJS, 214, 27
Mirouh, G. M. 2022, FrASS, 9, 952296
Mirouh, G. M., Reese, D. R., Faure, G., & Li, Y. 2021, MOBSTER-1 Virtual Conference: Stellar Variability as a Probe of Magnetic Fields in Massive Stars (online: MOBSTER-1) 49
Montgomery, M. H., Metcalfe, T. S., & Winget, D. E. 2003, MNRAS, 344, 657
Ong, J. M. J., & Basu, S. 2019, ApJ, 885, 26

Panda, S. K., Dhanpal, S., Murphy, S. J., Hanasoge, S., & Bedding, T. R. 2024, ApJ, 960, 94
Paxton, B., Smolec, R., Schwab, J., et al. 2019, ApJS, 243, 10
Stamford, P. A., & Watson, R. D. 1981, ApS&S, 77, 131
Stello, D., Cantiello, M., Fuller, J., et al. 2016, Natur, 529, 364
Townsend, R. H. D., & Teitler, S. A. 2013, MNRAS, 435, 3406
Verma, K., & Silva Aguirre, V. 2019, MNRAS, 489, 1850
Watson, R. D. 1988, ApS&S, 140, 255
White, T. R., Bedding, T. R., Stello, D., et al. 2011a, ApJ, 743, 161
White, T. R., Bedding, T. R., Stello, D., et al. 2011b, ApJL, 742, L3
White, T. R., Bedding, T. R., Gruberbauer, M., et al. 2012, ApJL, 751, L36
Wu, Y., Xiang, M., Zhao, G., et al. 2019, MNRAS, 484, 5315
Zinn, J. C., Stello, D., Elsworth, Y., et al. 2020, ApJS, 251, 23

Epilogue

Asteroseismology has become too vast to fit into a book of this length, so I apologize for what inevitably got left out, or failed to get the attention it deserved.[1] However, I wanted to keep things short and approachable, so I hope I can be forgiven.

In the Preface, I listed and thanked those who were here to welcome me to this field when I arrived. Let me end by thanking some of those who I've welcomed in my turn, and who have contributed - and are contributing - to making this the wonderful and welcoming discipline it is: Victor Silva Aguirre, Jeroen Audenaert, Warrick Ball, Keaton Bell, Earl Bellinger, Dominic Bowman, Lisa Bugnet, Siemen Burssens, Tiago Campante, Lindsay Carboneau, Ashley Chontos, Isabel Colman, Daniel Cotton, Guy Davies, Aliz Derekas, Lucas Eekhof, Patrick Gaulme, Charlotte Gehan, Sam Grunblatt, Oliver Hall, Rasmus Handberg, Emily Hatt, JJ Hermes, Dan Hey, Daniel Holdsworth, Marc Hon, Cole Johnston, Meridith Joyce, Thomas Kallinger, Christoffer Karoff, Rocio Kiman, Tanda Li, Yaguang Li, Chris Lindsay, Mikkel Lund, Mia Lundqvist, Savita Mathur, László Molnár, Joey Mombarg, Simon Murphy, Coralie Neiner, Martin Nielsen, Joel Ong, May Gade Pedersen, Filipe Pereira, Jean Perkins, Emese Plachy, Ben Pope, Angela Santos, Nicholas Saunders, Amalie Stokholm, Jamie Tayar, Andrew Tkachenko, Nathalie Themeßl, Timothy Van Reeth, and Tim White.

Our future is surely as bright as the objects we study.

[1] Convection, overshoot, and diffusion come to mind.

doi:10.1088/2514-3433/ae03a0ch9

www.ingramcontent.com/pod-product-compliance
Ingram Content Group UK Ltd.
Pitfield, Milton Keynes, MK11 3LW, UK
UKHW061825120926
8556IPUK00035BA/931
* 9 7 8 0 7 5 0 3 5 6 3 1 2 *